About the Author

Betsy Whittington

I am a registered Nurse who worked 28-years in Emergency Rooms. For about 14-years, I kept a journal. Journal writing was a method of coping with the tragic and sad things that happen to people—an outlet—a safe place to vent and share some very private and confidential matters.

Maybe sharing some of the stories will prevent or promote a better outcome for readers of all ages and from many backgrounds. Some may realize that they don't surfer alone and hearts may be touched with compassion for a fellow man or help them to understand better the emergency medical system and how it works and how it touches many lives.

In my book titled, "Choices have Consequences", Nancy Nurse begins her career in the E.R. as a new graduate. Her stories all come from an Emergency Room experience. These incidents and accidents didn't happen in a day, week, or even a year: but seen after many miles of walking those E.R. floors. Pain, tragedy, humor, grief and just people and the things they do and say when they are sick or injured.

Only about 8-10% of people that visit the Emergency Room are true emergencies—life or limb threatening. But until we have walked in their shoes—lack of ability to cope at the moment, pain, lack of medical knowledge, fear, lack of a private physician or one who will work a visit into their busy schedule—lack of common sense, money or even transportation. Some visit because of a need for the basics: food and shelter, well even companionship. There are lots of lonely people in this world., especially the elderly. If you are homeless and hungry the ER may be the only place to spend a few hours in a clean bed and be fed.

I am now a retired nurse who has written many poems and had a few published and one put to music. I am a Mother of five, Grandmother of six and Great Grandmother of four. I have written children's' stories who so far only the children in my family have enjoyed. Now I have the time to pull these out and have them submitted.

WAIT!

In the ER

Chapter Guide

The Emergency Room is has a character of it's own: There are the sounds—screams, moans, beeps, phones, radios, cries of anguish and monitors everywhere. The smells—the usual hospital antiseptic odor and bodily emissions of every description. The pace for the personnel anyway, is fast, hurried, rushed most of the time or that tense waiting for the next emergency to rush in those doors. To the patient waiting for labs and results or to feel better or to be admitted it can be painfully slow. The ER functions in military time so there is no confusion about am and pm. The cubby holes and separate treating spaces all have names and numbers.

The Doctors who Nancy worked with are as follows: Dr. Bee, Dr. Cooper, Dr. Stud Dooley, Dr. Remmey, Dr. Pickle and Dr. Playboy Stirella. Their characters will be revealed as the story unfolds.

There is a special wordplay among some of the health care providers:

Don't want to work syndrome—seeking compensation

Brain Dead—stupidity

Drug seeking behavior

MVC—motor vehicle collision

Fracture—it's broken

GI rounds—the Docs are at lunch

Coping mechanisms—rude, loud, nasty or the silent treatment

Pain—whatever the patient says it is

Drug reps—food providers

Holidays—special—we are always open

Inherent unpredictability—no day ever routine

Meaning of triage-sort out

Dumping ground—no space for patients

Language lingo: Example—pelvic exam—"who will slide down with me?"

Introduction

Emergency rooms across our nation care for the sick, the injured and the dying. They are busy, exciting and fast paced environments designed to care for you and me in our most urgent and devastating needs. Because choices do have consequences and because bad stuff does happen to good people, few of us escape illness or injury at some point in our lives. We live in a constantly changing social and cultural environment where stress and obesity are rampant. The family core values may be split and nobody seems to know "the normal" anymore. Maybe normal is just a setting on the dryer! Who lives in a normal family and what are the normal mealtimes and routines? Where is the balance in juggling our busy lives?

Emergency rooms have grown over the years to meet our needs on a 24-hour-365 day a year basis. Giving emergency care requires knowledge, wisdom, dedication, compassion, fast thinking and critical decision making skills; Add teamwork, confidence, competence and ability to that. And, I forgot to mention endurance and patience.

This story takes Nancy Nurse from a student nurse and ends with her end after retirement. The location if you local Emergency Room and includes many stories of the sick and dying patient and tells how Nancy copes and handles the daily pressures as the sick were cared for.

There have been many battles won, many victories, lives saved and limbs restored. Let's focus on the positive side. Roy was a neighbor of my parents and came to the ER from the local basketball game in full cardiac arrest. Quick treatment and a shock to the heart revived him and he has lived a good quality of life the past 10-years since the incident. Then I remember the motorcycle guy—he came in with a mangled leg and survived surgery and is now riding a motorcycle with one leg. Nancy still bears the fingernail imprints of his hanging on to life and her. There was the young gentleman with the gunshot wound to his chest—Nancy remember a whole as big as her fist. He's alive and well. The long evening after the tractor trailer accident—the husband died instantly and his wife who was riding along had a terrible crush injury to her leg. She has her life and limb. How many young men are living full lives after having the clot buster drug and heart muscle saved after the big M. I.?

How many wounds have been repaired with beautiful results and how many broken bones have been returned to full function? How many children with ear pain and infections have been treated and comforted? How many patients with viral illness and gastrointestinal symptoms have been relieved and hydrated with intravenous fluids and medications?

How many patients have been seen, diagnosed, treated and cured? Certainly there is only a small percent with tragic outcomes—but those are the ones remembered. How many patients with high blood pressure have been saved from strokes and paralysis? How many pelvic infections have been treated saving the child bearing function? How many people living to a productive, ripe old age because of treatment? The doors remain open and the sick and injured continue to seek help.

Chapter 1

Nancy Nurse

Florence Nightingale instills fairy tale ideas into student nurses. They arrive each morning with faces scrubbed bright—only a touch of makeup and with white caps perched on their heads. The dresses are two inches below the knee and shoes are scrubbed with whitest white ready to save the world. How long does disillusion take to break into their heroic dreams: the first unkind response, the bed pan running over and smeared about or an untimely death of tender age.

Nurse Nancy came through training still untainted and seeking more experience and education. "I know so little—medicine is so complex." Nancy gets and introduction into mankind's ill in the office of Dr. Ned, an old general practitioner. Open office visits and even house calls fill the days and offer family treatment, penicillin and pain relief. Patients instead of paperwork seem to be the priority. Dr. Ned suddenly died and Nancy is hired at the local hospital.

She begins her hospital nursing career as a float nurse: each morning taking an assignment in the specialty area that needs help because of sick calls or heavy patient census. Viral signs, baths, treatments, meals, medicines, doctors, orders and paperwork and charting and more paperwork and telephone calls, lifting in the bed, lifting out of the bed for x-ray or testing fills the hours. When is this comforting and teaching supposed to take place?

The days go quickly and lunch break is short. Even in those thick white soles my tootsies ache. Orthopedics, traction, casts, check the circulation—relieve the pain and they still need baths and linens changed, new tubes and change the catheter bag. Nancy Nurse, I need something for pain—no, not the shot—call the Dr. for a pill. The patient care is still a learning experience, but how do I read these doctors' orders? Louise, can you help me out? Dr. Lewis even asked, "What did I write here?" Charting, calling, scheduling tests, rounds with the Docs: Harry Jones is going down the tubes. The pull continues throughout the day: bedside care, charting, document, document, document. If it isn't documented it isn't done. In school we had two patients a day—not eight or ten.

Nancy works and Nancy learns from O to Psychiatry to Critical Care. She floats about the different units and works on her skills and signs up for continuing education classes. And then one day Carol in the ER threatens miscarriage so Nancy is floated down to cover her shift. Hey wait, "I never made the ER rotation in school!"

It's evening shift and this place seems strange. Billy Jo, Scott and Tome: three male nurses in one place. What do I do now? Go to room number 3 and set up a suture tray. I remember sterile technique but patient George has an ugly wound from a chain saw encounter. I don't have time to feel sick. "Take the vital signs." states Dr. Bee. How does he look? Is the bleeding controlled? The evening is filled with sick children, fevers, patients with congestions, back pain, that wheezer is a regular—give the usual! Regular, what's the usual? Mix an aminophylline drip. Let's see, how much—what rate? Nancy, you'll do fine, you have what it take states Dr. Bee. Boy, what a night, I'm glad someone thinks so. I don't remember these scenes in my heroic dreams of nursing the sick.

Chapter 2

Tragic Horror

Do you choose to fly? Everything has risks and reward and consequences. The radio system comes alive. The voices are excited. Do you copy? There's a plane crash! A small business plane carrying six passengers and the pilot and the six are dead! Set up the trauma room for one. We have a list of the victims, the pilot is the only survivor. Nancy, get on the phone to relatives. Hello, there has been an accident with the plane. Could you please come to the Emergency Room? Hey, two of these people are my neighbors. I know Mr. Miller. Not one of the relatives asked "the question" over the phone. Yes, I can greet each family member and tell them their loved one is dead The pilot will be ok—physically, I think to myself, maybe.

Where is the clergy? It's too early and no one is on call either. Let's see, death—which class covered this? I remember, sudden death, which was explained as acute pain—like a spike in the eye. Don't use other words: just state clearly—he died—he's dead. Where are the tissues? I didn't know this morning that I would see so much grief, tears, fists pounding. I can't help if my own tears get in the way. I'm sorry, I'm sorry. What little comforts we have to offer, a hug, a tissue. Nancy will later learn of the power of "The Comforter". Am I supposed to feel this helplessness? Nancy Nurse drags home to her family with tears still slowing; the emotional work is very tiring. As she tries to fall asleep the scenes keep repeating in her head, I'm sorry, I'm sorry.

That radio again, maybe it will be a signal 2, no transport. The drowning of a two-year old and they are fifteen minutes out and cannot intubate? We retrieved her from an outhouse, an outdoor

John. Grandma was watching her and she seems pretty loaded. Grandma is drunk and the baby is dead. Those soft blonde curls in a blue, still lifeless body, feces still clinging to her small hands and feet. Nancy, come out, it's ok to cry—but what a needless death. Her caretaker is drunk and it is too late. We can do nothing but put a tag on that tiny toe and wrap her in a plastic sheet and go home with tears and clinging memories of life's tragedies Why—why—why? One day Nancy will realize that all choices have consequences, but that God is always in control...

Tragedies seem to be a way of life. The 62-year old man in bed 2 is very confused and has left sided weakness. His CT scan reveals a bleed in the brain. Suction, call air care—get him to a neurosurgeon?

That man with chest pain cannot have a lift threatening dissecting aortic aneurysm. We have too many other critical patients now. Could we please space out those emergencies a little?

The 50-year old with chest pain said his pain is better The EKG looks normal. Are those coronary spasms—V-Tach, thump to the chest, ventricular fibrillation, Zap, Zap, Zap, shock times 3. He looks up—"I don't feel good." I'm so glad you can tell me that you don't feel good.

The patient in the parking lot is having a seizure in the wheelchair. The feet are on the ground and I can't move the chair. Do I have here for help or until the seizing stops? Decisions, decisions and a life hangs in the balance.

The only Rx Mr. Jones get is a toe tag, a shroud, and the sheet over his head. Was death on his agenda for today? Would his yesterday have been any different if he had known? Did he say a special word of goodbye this morning?

Sallie got up to go to the bathroom and IV line got caught and dislodged. The IV catheter is still in place with blood going everywhere. I guess she didn't see the call light. It looks like a bloody battle has happened here. What a mess! Now what is her hemoglobin?

Where do you live? "All over" is Ned's reply. He is wearing 13 pieces of silver jewelry, bilateral knee splints, bilateral ankle splints and is on the phone calling ads about a live in. Hi said we stole his watch because he has an empty box? He has an empty box alright, but not in his hand.

Nancy was called in to work at 4 am. There are two codes going on with one being a near drowning. His blood alcohol is 370 which means that is 3.7 times legally drunk. Nancy was showered from the patient immediately upon arrival with pink, frothy sputum. Is he an organ donor? Student Nurse Sue is so new that when she pulled back the sheet to insert the Foley catheter that she gasped at the male part. Yes, that's normal male anatomy. But Nancy remembers that all is not normal—at least not the last patient she tried to cath unsuccessfully—no labia. "I used to be a man, the MD said he cut it off, tucked it in around the edges. Looks pretty neat?" OK.

Some times it is just too late—too late. This child has the heart muscle lacerated, no breathing, no heart beating. My son, my son—the child is dead. Oh, his poor brother. His brother was using a bush hog type weed eater that hit a stump and kicked back into his brother's chest. Mothers tears, my tears, everybody's tears are everywhere. The tragedy stalks throughout the day—voices quiet—

the care gentle—something hangs in the air. Grief, heartache, empathy. Why? Why? He was such a good boy! Years later Nancy still will cry, as memories hang vividly in her mind. Some she will never forget

"I feel chest pressure, like something setting on my chest and I am short of breath." Nancy, put the oxygen cannula in place. What does the heart monitor show? Start an IV line, draw the labs and get the nitroglycerin. The patient is too quiet, skin a little moist and pale. She looks into my eyes and says, "I'm going to go now." and proceeded to roll her eyes back, stop breathing and die despite the Code 90 call, the CPR, the shocking, pounding and drugs to stimulate a heartbeat. "I'm going to go now." The monitor shows a heartbeat but the patient has no pules: electro-mechanical disassociation— EMD. Her family doesn't understand. She walked in here with a heartbeat and breathing. She's only 60-years old. "We've been at a convention dinner." Death know no bounds—old or young—home or away—suffering or sudden. The enemy lurks, waiting to attack it's prey.

"Good morning Billy Jo. Scott—this place is quiet today! Patient only in beds, 3, 5 and 9. Oh, the patient in bed 9 is just sleeping one off, his blood alcohol showed that he was two and one half times legally drunk. We can't find anyone to come for him." Time to restock—is the crash cart locked? I'll put oxygen cannulas at every bedside, bed pans, tissues. More needles for the IV trays, bandages, tape. Maybe everyone will be careful today, drive safely, eat healthy and stay well. No accident today—no sickness—no death. I haven't seen one of those days yet but there is hope. The radio goes off and there has been an accident, a head on collision! Wait night shift, don't go home. There are at least 8 patients with injuries and the ETA (expected time of arrival) is 8-10 minutes. The tension mounts as IV lines are set and the trauma packs readied.

Patient #1—an 18-year old female with multiple trauma and unresponsive.

Patient #2—vital signs are stable and there are multiple lacerations and abrasions

Patient #3—No pulse and no blood pressure. Dr. Remmey pronounced him dead at the door.

Only 30-minutes later—did someone see a hand move as they passed that curtained cubicle. Two patients have been sent to the operating room and wound repair is in progress on another. There are two still in X-ray. You must be seeing things, he was pronounced at the door. There is a faint pulse—call a code 90—get Dr. Remmey. Start an IV line. Attach the heart monitor and get the oxygen. Well, George Irving comes back to visit the place of his resurrection. I am glad he is not bitter or angry—just thankful he is able to walk through those doors. How could it happen? If you look at any dead person long enough they seem to move or take a breath.

Another day, another dollar, another death. I don't remember that part of the story. Another car accident and the young mother just took the kids out for an ice cream. The children still have ice cream on their faces along with the cuts and scrapes and broken bones. "Where's my Mommy?" A drunk driver hit their car head on and Mommy is dead! Grandmas is called to the ER and goes from child to child. She is taken to a private space and told of the death of her daughter. She pounds her fists into the wall and the grief engulfs us all.

Chapter 3

Character of the Emergency Room

Dollie and Earl are two of the most pitiful examples of humanity I have every encountered. Dollie is wiry, whiny, anxious and has every complaint in the book. She is shifted in the medial system from office treatments, multiple workups and tests galore, but doesn't get the most important treatment. She is in dire need of psychiatric help. So much of our anxiety and thought patterns make us physically sick. She has so many mental needs and torments that are not addressed. Too many times the patients fall through the cracks when the Doctor doesn't take the time to get to the real issues. In the ER they just want to see them and street them—just move them through the system. Does anyone have time to care? Dollie has brittle, over-permed hair, a grating voice and is very restless. She refuses and then requests intravenous fluids and lab studies. Her fingernails are heavy yellow with nicotine stains and she decided to see the Doctor too, since she is here with Earl.

Earl has a gray color and wet, clammy skin. He is unshaven with greasy stringy wisps of color treated hair. Maybe he is a cardiac cripple and some rehab and teaching could turn around his fearful, painful existence. Is the chest pain real or does he just want the morphine. Is his real need attention or encouragement?

Babies can be so sweet and soft and also so stinking. This little girl has a big poop and smells of soured milk. I know that a clean bottom and warm bath would make her feel better and not so cranky. I can only try to gently clean with a washcloth and pray for some supervision for that little Mother.

The next patient is an ER frequent flyer. She has a decreased level of her favorite psychiatric drugs and has had "abdominal pain for 10-years". She states, "I have the shakes and am fixin' to die.". The MD said, under his breath, "That would make things much simpler."

Is this save the fossil foundation? A cruel uptake on all the nursing home patients today. They all need total workups and maybe an admission but they are not allowed to die in the nursing home so they are sent to for that final round of sticks and drips. Where is death with dignity? We are all born to die and sometimes a celestial discharge seems to be such welcome relief.

There is a nest of buzzards on the hospital roof and pecking at the windows. "Stop, you are making the patients nervous."

Those paramedic students, "Let's have a good day!". That means they want to see lots of accidents and sick people. Does that mean that someone has to have a really bad day? Can I watch them bleed?

The lady present at the Triage desk—"Can we sort this out?" "You want the cure for what?" "I can't stop laughing!" states the middle aged black lady with loud laughter, that's contagious—now the whole waiting room is in an uproar.

The next room reveals one nasty fracture—connected to one nasty lady. How can they be so rude when we are trying so hard to help them? She is "going to sue everybody involved!" Does it matter that she is fat and clumsy and tripped on the curb.

Tip of the day—hair spray will remove and dissolve blood stains.

He was struck by a car and calling it aggravated assault. Can you give him oxygen and transfer him to the trauma center. All thirty of the X-rays are negative and his family called the local TV channel. "Why am I working here?" The whole duty of man is to love and serve God ad if I wasn't serving him I couldn't take the abuse.

It is 7 am and Mommy looks very drunk. The 8-year old child states, "Mommy drinks all the time." There has been previous HRS involvement. What heartaches and complications for this child.

It's a weird morning and the moon is not even full. It's the weekend, Saturday , and the patients are coming in droves. Don't they have anything better to do with the weekend? Ok, I know, they come for many reasons and only one is that they need emergency treatment now.

"You will be back in a moment? Is that a hospital moment— about 45-minutes?"

The ugly little lady sits at the triage desk. "That's why I don't like to come to the hospital—that blood pressure cuff is so tight." She is so ugly and the wrinkles are so deep on her face they look like craters her arms are so long they reach her calves and her face, the shape of a mule There is a medical condition behind the look. She is so short of breath but she isn't concerned about the breathing. "Don't bother with that—just look at my leg." it is red, swollen and hot to touch. "I'm sorry." I struggle to remember that sickness and pain are part of our fallen world.

"I just need drugs." Give me the administrator. "You are threatening my well-being."

There is no patient like an old nurse, "Did you pull back on that needle, did you check my arm band? I need some Demerol and I can't move—just cut the clothes off." I'm getting to be an old nurse and I give good care, but it takes time, give me a minute.

What are the results of the pap smear and cultures? No GC, no sexually transmitted diseases. "Are you sure?"

Ugh—He was claustrophobic and jumped off a tram and amputated the foot.

The family states, "in three weeks his only words have been—swallow is a bird. "

I need a face mask and where is the air freshener? Has he slept with a pig or a pack of dogs? His is so dirty and smelly with a large hematoma on that shaved head. Do we just clean the head wound or find a shower for the whole body? We really need a robo wash—where you slide the stretcher through.

The chaos that went on here today would blow the mind of St. Peter watching from above. The little man all diaphoretic and with chest pain sits patiently and quietly and watches. The little couple in their 80's are hovering over each other. He is on his cane, wearing a red shirt and sporting a little mustache, "I can't leave her here, she don't eat enough." She said, "I didn't pass out, I am just tired from worrying over him."

No death with dignity, no respect for the elderly. The 94-year old has terminal cancer and a 104.6 ° fever. He has a large gaping sore on the left hip and now we have to stick tubes in every orifice and figure out what is wrong with him? Can't we please find him a private room and let his family surround him and make him as comfortable as possible?

The 46-year old male says, "Help me, I need oxygen, give me air!" and to his 10-year old son, "Justin, I love you." He is making me feel short of breath. The sweat is pouring off his skin and he is struggling with eerie breath and that bluish color is overtaking his body. He has a collapsed lung. Dr. Bee takes that silver metal pole and pushes it through the ribs as Nancy tries to calm the Doctor and the patient. Those pleading eyes look at me as he reads the name badge. "Nancy, help me." The chest tube is connected and the water starts to bubble in the closed chest drainage system and patient starts to breathe deeply and his skin color starts to pink up and sweat stops pouring. As I affix the dressing over the sutures holding the chest tube in place I am exhausted. The emotional work can be overwhelming.

It's 7 am and a lady runs in the door screaming, "I need help getting Dad from the car!" The car sets in the Emergency entrance with a gentleman half on the ground and a wheelchair by the door. The daughter is crying her eyes out. How did she ever get him into the car? Sometimes, or even when in doubt, just call 911. The will send a bunch of healthy young trained and knowledgeable people to help you. And sometimes minutes do make a big difference.

Some days IV sticks are easy and sometimes the veins seem to fly away. Some patients make it difficult to find the best vein. "Don't stick here. " "Why did I let you give that medicine?" "My Mother died young of a drug reaction." (Maybe the source of her anxiety.) "This is the only vein they ever get." "Now use a butterfly." (We haven't used butterfly needles in about 20-years.)

A 25-year old is yelling, "They are going to kill me? (the Nurses) His speech is impaired, "I not retarded, this was caused by my Father's abuse. My belly pain—it's getting worse."

Rx request—the 39-year old with back pain and groggy speech states, "I want Demerol." as she wadded the chart and discharge orders and threw them at me. "And I want a muscle relaxer!"

33 is too young to die. I hope the clot buster saves his heart muscle.

Screaming Ruth, I told her she was ready to go home and she starts cussing and fussing. "I'm dying and my arm is falling off and I can't use the bathroom. Bring me a bedpan." Is she lonely or afraid or what kind of environment does she not want to go back too?

What are the sounds in an ER? Zap, zap, zap. Monitors beeping and then an unusual alarm that sounds like Big Ben. Did someone plant a bomb? It wouldn't be the first bomb scare here.

Let me calculate the medication dose. 60 mg from a 2 cc vial of 125 mg. Can't the Dr. just order on or one-half vial instead of making me remember my algebra?

All the treatment rooms are full and now we are filling up that dark hallway. At least we can begin to order their treatments. Why don't we have enough ER spaces for all the sick people? The town has too many hotel rooms and restaurants all over. Build the spaces and the people will come. Maybe a little competition will help waiting times and the treatment options.

The secretaries spent all morning entering data for the log in the computer and it didn't store any of it, while the nurses ran to the lab, answered phones and took care of patient orders as well as the patient. The tech is hiding out in the med room eating sunflower seeds.

We still do good work! The treatment was fine until you got the bill. $847 for that tiny kidney stone. Wow. And $1,600 for a sprained ankle, oh X-rays were included.

We have seventeen beds full of sick people and four new ones from a bus accident. They all have neck pain, can we sue the bus company?

We have ten people waiting to the Dr. and a man sitting in a wheelchair screaming, "I need a pain shot now! I was here two days ago with a kidney stone and the urologist saw me today." "I can't tell you my name." "I don't need a chart." "I am going to throw up." "My Doctor said he wound meet me here." "I need a bed, now!" "No, you can't take my blood pressure and I don't have a temperature." All the other patients look on as if you are beating the poor man. I do need a name to find a Doctor to treat him.

We have had six patients with back pain today. They are all young males on workers' compensation. Each would like six-weeks off work with pay. Maybe the weather will improve by then.

Lip laceration forehead laceration, chin laceration—are we having a quilting bee today?

The little guy with the MI is pretty sick. His upset wife feels dizzy. She sat down and ate something and is still dizzy. Let's check her vital signs, she doesn't look so good. Her heart rate is only 30. Let's bring in a stretcher, get a monitor, the O2—start the drips. Can we give them side by side coronary care beds? How sad!

The motorcycle victims struck by the vehicle of the drunk driver look bad. He may lose that leg and her left leg looks even worse. Later that evening we learned that they both had amputations of the left legs.

Chapter 4

The Sick You Will Always Remember

There are sick people everywhere. The diabetic in room 1 has an infected stump. He said the toes were the first to go and then the leg and the other leg is not doing much better. Will he have to have more of the stump chopped off? It takes a lot of faith just to face the next day sometime.

The little lady in bed 8 is 101-years old and has a sharp tongue "It's all foolishness." she states. She has a terribly infected leg and doesn't wasn't any treatment but the daughter insists and they are battling it out. She said the she is old enough to take care of herself. Sometimes it is not the rebellious teenagers that create the problems.

The guy in bed 2 is in the area to visit his two granddaughters. He had a pacemaker two weeks ago and now is having a stroke and is unable to speak.

We have six patients en route from their job site. They inhaled something and all are having respiratory difficulties. They inhaled floor stripper, which caused a carbon monoxide buildup. Which of the six finally realized there was a problem or did they all suddenly have trouble breathing?

We have two young girls at the same time with new onset seizures? What is the cause—low blood sugar, video games or a deficit in the brain? There are lots of issues to sort out and lots of anxiety for two sets of parents.

The freckled, thin 23-year old male in the next treatment cubicle has all his fingertips rotting off! He has a history of Raynaud's disease and Scleroderma. The fibrous tissue turns to scar tissue instead of normal tissue. His future certainly appears to be full of heartache and pain and suffering. Why! Today I will eat, drink and be merry and very thankful for my blessings.

The 52-year old man in room 5 is too quiet, too pale. "Just that knife like squeezing pain in the center of my chest." "I'm healthy as a horse." Maybe you were—but you're having the big one. Start the oxygen, intravenous lines time 3, EKG, chest X-ray—make it portable, nitroglycerin-sublingual and hang a drip. He needs the clot buster drug. (We won't tell him this stuff costs $4,000 a shot or that there is no battery in the monitor or that the IV line refused to thread and that the repeat sticks put him in danger of bleeding, the EKG didn't get the page and there was no aspirin in the pharmacy drawer.) I just wish we weren't so prone to human error. But the thrombolytic save his life plus heart muscle so I don't think he cares. His blood pressure is dropping, decrease the nitro drip and lower his head. Those nasty reperfusion beats make my heart skip.

I feel like I am stranded in this war zone. Whoever constructed an Emergency Room with no windows. We don't know the weather—unless we get a bunch of motor vehicle accidents at once—10-minutes post rain. We don't even know if it's night of day without the military time. We opened our door, turned on the lights and they have never been turned off—24 hours a day, 365 days a year including all weekends and holidays. The sick never stop coming!

As I walk to the triage desk brightly at 7 am with my coffee cup in hand this gentleman is humped over with his legs apart. "You got to do something with the hemorrhoids." I didn't walk like that when I had a baby. Nurse, nurse, where is your compassion? You only know what you immediately see. The stories and lives can be so complicated and full of tragedy and sadness.

There is a white rental car in front of the ER door with the key in the ignition and the door open. There is no one around—no dead body. Oh, that must be the anaphylactic reaction that went in for treatment about 10-minutes ago. Seconds count when you are having trouble breathing.

"Bad food, I can't breathe." He has a low blood pressure—96/44 and only a 41 pulse with a respiratory rate of 28. He has loud wheezes and his lips are edematous and white. Tech, find him a bed (a spot in the acute treatment area). What is you birthdate? (as he is heading toward death). In the treatment area his skin is beet red and he has hives all over the trunk. Give the epi, Benadryl, Solu-Medrol 125 mg states the Doctor. Get an updraft treatment with Proventil. Start the oxygen and get IV lines, now. He is having a sever allergic reaction to a nut. Intubate, he's no breathing. He's too young to die, he walked in. The struggle is to keep him breathing.

My next patient fell 30-feet and landed on his feet. He has bones broken from the heels up. It took 14 X-ray pictures to find them all. He even has multiple pelvis fractures. What an impact!

The guy with check pain refused the nitroglycerin. That's our first line drug. "Nope, that's what they blow up bridges with."

The man with acute chest pain in bed 1 is to be admitted for observation to rule out an MI after the Doctor get a view of the chest x-ray. Whoops, he has a pneumothorax, 75% of his lung is down, collapsed. He needs chest tubes, not nitroglycerin.

The next patient appears at the triage desk and states, "I'm bleeding." He is 30-years old and looks dead with that awful gray, white color. He had a history of a platelet disorder and he doesn't mention his HIV. He has a hemorrhage in the left eye and "feels clot of blood in the throat". His pulse is up, 132 and his blood pressure is down, 90/50. He won't be long for this world and I can just hope that his heart is prepared for the next.

That is a hateful little 75-year old man with a string tied around his neck. I hope he doesn't hang while waiting for x-rays. "I've been her an hour already." How did he get those rib fractures? And what about the crazy lady who is ill (because of silicone breast implants she received 12-years ago) and she refuses all tests and wants to leave—but she came for help and she so needs a psych admission. And the black guy in room 11 is afraid he will kill his kids and skips the joint while the Barker Act is being processed. Call the sheriff! The lady next door to him had a panic attack en-route for a sonogram and is crying and praying loudly. The sick you will always have with you, I repeat in my head. The sick you will always have with you.

How bad do you feel? We will put you in a cold treatment room with an open gown and the Doctor will see you in a couple of hours. We can't go any faster. We need more Doctors, more nurses, more secretaries and more treatment spaces. Then we need more x-ray and lab technicians. Sometimes it seems we only get two minutes with the patient and it takes 20 to record and make computer entries. But if it's not recorded it's not done per the lawyers.

I looked up to see a 500 pound man filling the doorway. "My tummy hurts." he states. "All of it." Let's find him a treatment space. The Doctor's on a GI. consult. I can never say the Doctor is at lunch with all of these sick people waiting. "I am here dying, does he have to eat?" Sew one up, tap the fluid in a knee, shock a heart: convince the man with the acute heart attack that he is sick and needs to stay in the hospital and the guy in the next bed he is well enough to return home—all in a days work!

Typical ER nurses have strong personalities: assertive and aggressive. Do you know what happens when you put two bulls in one pasture? You hear a lot of bellowing and then they lock horns. There are different management styles and often more than one way to get the job done and with a good outcome. Nancy Nurse has back problems or maybe just PMS but that can't get in the way. Only the neurosurgeon gets away with his rudeness—and to the patient.

A 23-year old sits staring into space. He is oriented to name only: labile—crying. He was called in by med com as an emotional disorder but has a blood sugar of 23 (normal is 80-120). We give Dextrose 50 IV and then lunch. He is out of here and "feels great". He is a diabetic who went too low—we do good work!

What is on our agenda for the moment: One fish bone in the throat, a frail little nursing home patient with gross hematuria (blood in the urine); a tourist with a fat swollen ankle; a simple rash—or maybe not so simple. She looks very ill and the joints are sore. A patient who is Jehovah Witness and has had vaginal bleeding for 3-weeks and her hemoglobin is drastically low—only 5.1. They only want comfort measures—no blood. She is too young to die! The next patient has left flank pain and diaphoretic—Let's get the heavy jacket off—r/o kidney stone? The next patient said he was beat in the face with a pipe—he's so rude, maybe we know why.

Just because it's interesting—she has a bug in the ear—it's a roach and she is grossed out; he caught the big one, a fish hook in the buttocks; a vibrator in the rectum, "don't remove, just change the batteries". It takes all kinds to make the world go round—I guess.

Tragic things: a heavy black lady who didn't want to wait and crawled under a train—she had both legs cut off.

Difficult removals: the chef came in with a steak cubing machine attached to his hand. And the young man has a nail through the sneaker, the sock and his foot.

Three Spanish speaking children and their mother are all restrained and immobilized on backboards after the auto accident. Everyone is frightened, but there seems to be only small bleeding wounds. It's the noise level that sets all tempers on edge—and I know it can't be helping the lady in the next bed with an anxiety attack.

The child is ok, maybe a little dehydrated. The Mother may not survive watching the IV start.

The black lady is calling for a ride home after her treatment. She said, "think about having kids, don't have kids, choke them! I have 12 sons and 2 daughters and no one can come for me".

In room 3 is a 20-year old black girl named Georgie and her head is shaved and she has multiple small wounds to the chest—all self inflicted. She overdosed on Zoloft—to feel better. She has been given the charcoal to absorb the medicine left in her stomach and now has vomited the black stuff all over and has diarrhea which runs onto the floor from her clothes and the bed. That is a yuck smell and difficult cleanup before coffee. She said she has "girl problems".

Little Harry from the Nursing Home is spitting up horrible

goobers. "Whatever you do is ok—I can't live too much longer anyway." "Pneumonia—yeah—that would be a nice way to go."

"I didn't ask to come here, I just said the F word at the wrong time and three black guys came after me. They could have left me lying on the pavement in a pool of blood with a broken nose and my face all cut up." Sorry you must stay here until you are no longer legally drunk—a blood alcohol of less than 1.0.

This wheezing child needs treatment. "Well he stopped breathing the last time this happened." Let's see—we already have a COPD'er in front of the desk on oxygen and all the treatment spaces are full. That's the same gentleman that hit Nurse Brendy in the jaw the last time he was here. He couldn't breathe and he didn't like her attitude. Let's move somebody!

That little 90-year old has slurred speech and a slow heart rate. He is probably having a TIA—transient ischemic attack. Old age just got more difficult for him.

Old and young, rich and poor, makes no matter, they come through those ER doors. Left leg cellulites, right leg cellulites—the two are in rooms side by side. Start the IV's and mix up Ancef—2 grams for each.

There is a 8 year old man in room 1 with an asthma history. He has wheezing throughout the lungs and is in congestive heart failure. He does have a history too of atrial fibrillation. The IV Lasix is to remove the fluid but he only wants me to call his Doctor, "He's the best!". The best Doctor is 5,000 miles away and can't see your EKG or listen to your breath sounds.

The 51-year old in room 5 said his right chest hurts and the ribs and gets worse when he lays down. "I was dead a week ago—an employee put heroin in my coffee and I had cardiac arrest." Somebody kicked me and started CPR and called 911. You are one lucky guy. A live man with a sore chest—some Advil should help that.

The quiet sadness hanging over as we enter the ER for the 7 am shift. There is a 27-year old male just pronounced dead. He has a deaf wife and four small daughters: ages 6, 8, 10 and 13. The Mother needs help from the thirteen year old just to know what is happening and to help make some arrangements. Three little girls just need to be hugged, held—Dad is ok. I know your heart hurts.

And in the next cubicle is a 30-year old diabetic. " Yes," he said, "I take insulin twice a day, sometimes—when I can afford it." He has a cold, white left foot and a big blue toe plus edema of the entire left leg with a rash. He has sever peripheral artery disease plus the diabetes or because of the diabetes. He may lose the toe, foot or the leg! And in the next room another insulin reaction. The world seems to be full of diabetics and lot of them with little understanding of their disease and its side effects and poor control. You take insulin and then eat within thirty minutes because it acts on your food. We are digging our graves with our teeth and changing our lifestyle is so difficult with the emphasis our culture has on food and so much of it unhealthy food. The price tag for obesity is staggering.

The paramedic states, "27-year old female, patient unresponsive, airway blockage and no blood pressure." She has epinephrine and Benadryl on board, the patient is responding. She appears to have an acute allergic reaction with red skin from head to toe. Next time she comes in I hope she wears something easier to remove than this jumpsuit. She almost didn't have a tomorrow or a next time for anything thinks Nancy.

The 27-year old male in bed 8 has no arms or legs—a Thalidomide baby. He is having abdominal pain and flops on the bed like a fish to move himself. And I complained this morning that I was having a bad hair day. I can walk, talk, feed myself, jump, dance, sing or at least make a joyful noise. I go to the bathroom with all bodily functions intact—no plastic bags to catch excretions. I don't need a shot or two of insulin daily to live. I have no tube down my throat to breathe for me or a feeding tube into my stomach from which some luscious white stuff drips to nourish me and hopefully no tumors growing inside to zap my normal development. I'm not curled up in a little ball with bedsores on my behind and limbs contracted so they can't straighten and smelling like old pee. Existing and living, quality verse quantity. Thank you Lord, I have every reason to count my many blessings.

Could I cope with the issues that Mrs. Allen does each day. Her husband had a stroke 10-years ago and looks like skin stretched over bones with a tube to feed him and a tube to urinate. He doesn't speak and barely moves but she takes excellent care of him. Such devotion—but the way she beats him on the back to loosen secretion, I wonder.

Sadie's son-in-law was in an auto accident. When he came in to the Emergency room nurse Patty raised the head of his bed. Did this cause the spinal cord problem that caused him to be paralyzed from the mid chest down? After a year in rehab he came home to his beautiful young wife and young daughter. He was so angry and mean that she finally divorced him and he retaliated by stalking and shooting and killing his beloved wife and spends his life in a wheelchair in prison. Sadie has her granddaughter to raise. What if—the first responders had immobilized his neck and what if my young co-worker had not moved his head? I don't know. I don't know. Our mistakes can cost lives. It is so scary sometimes—pushing powerful drugs into IV lines and patients circulatory systems. This stuff costs $4,000 bucks a dose. What if I drop the vial? And then shocking 360 watts of electricity into someone's body and seeing them flop on the bed or sorting through the symptoms and working to ding the right puzzle parts to help the Doctor make a correct diagnosis with proper treatment.

A wrong medicine on a wrong chart. A dose ordered inconsistent with life. Dr. Jeri wrote an insulin order on the wrong chart and the nurse carried out the order. Quick, give D50 before her blood sugar drops too low and we will need to observe her the rest of the night. If it was my mother, I would be understanding. The five rights must always be in the back of our head, the right medicine in the right does to the right patient at the right time and in the right mode of admission. Oh, how I pray the my mistakes will never be a fatal one.

Katee is a nine-month old playful red haired little cherub. Her Mother said she cried all night—after she had fed her a plate of spaghetti. So—I would have a tummy ache too—don't feed a nine-month old so much spaghetti—some things are very hard for young mothers.

George is a 30-year old that awoke at 4 am with chest pain. He went to a 24-hour pharmacy to get Mylanta. He is in the ER at 8 am with the big MI! He is quiet, rubbing his chest—just heavy here. Can we save his life and can we save the heart muscle. Dull, heavy or tight pain in the middle of the chest that doesn't go away and especially if associated with nausea or sweats should be evaluated immediately. Those are the people we keep the night lights burning for.

Elaine is at the triage desk and just wants to make an appointment. Could you recommend an internist? Her blood pressure is out the top—189/116. Her pulse rate is too fast—108 and she has a fever of 100.7° and is very short of breath and has chest pain. No, Elaine we won't make you wait for an appointment with an internist. She is a mess and ends up having pneumonia and angina. Again—the people we keep the midnight oil burning for.

Tootsie is 90-years old and a frail little thing with floppy boobs hanging below the waist. I wonder if she went braless as a teenager? Now, I know why I don't need big ones. I had to lift on up and flop it across the shoulder to attach the electrodes. She states, "I might smoke, but I don't inhale and besides that has nothing to do the breathing." Choices have consequences Tootsie: we reap what we sew and in a different season. Your 90-years prove that you have had a prolonged season.

That leg looks bad. The 31-year old German male has a bad cellulites of the left foot. He has edema and red streaks up the groin. His temperature is 103.9°. If he had waited any longer to treat that staph wound he may have lost a limb—or his life. Those who should come to the ER wait, or don't come and those who could easily treat themselves with some common sense, crowd the corridors.

Fred didn't like hanging out in the ER and signed out AMA— against medial advice. The Dr. wanted to admit him to the hospital. He slipped in the doorway as he was leaving and now has a bad fracture of the left ankle—Oh no! You can't sustain an injury in the ER door. Accidents do happen—bad luck? I hope it wasn't from a wet floor.

A bird's eye view of a panic attack: "I'm on the edge, I'm going to faint. I need a cold rage. Get my blood pressure down." Look at those spastic jerking movements. "I can't stop my leg—there it goes again."

Caralee took two blood pressure pills instead of her 2 vitamins. Cardizem DM is a long acting drug too. He blood pressure of 84/50 dropped to 60/40 when she stood. No wonder she feels dizzy. I know the hospital and nurses make drug errors but I wonder nationwide ho many home errors are made daily, especially little old people who are on lots of meds. We are taught to check 3 times and to make sure we have the right patient, the right drug with the right mode of administration and at the right time.

The lady in bed 2 is a known diabetic. She has a blood sugar of 1,792 and just had a seizure.. (The normal is 80-100) She is on vacation and left her insulin at home. No—please, you don't get a vacation from your diabetes or the medicines that you normally take daily. Will she suffer permanent consequences?

The male diabetic in room 4 awoke unable to speak. He wife said, "OK after forty-years I can have my say." He has a droopy side of the face and is unable to raise both arms and a new CVA—stroke.

We are waiting for the medical examiner for bed 7. The patient is a 50-year old black male—dead. He has four foster kids. Who will look after the children?

In the next cubicle is a young, handsome, dark, dirty haired man. He has feet problems—dirty, cracked and red. His affect is strange. Is he a drug burnout? No address—smelly. I don't think I want to be a bag lady and live on the street where there is no bed to sleep in or no regular food. Does he remember the temptation of that first joint—peer pressure to be part of the crowd? No one ever intends to use to the point of addiction. The deception of the evil one!

The forty-seven year old in bed 2 came in complaining of trouble breathing. She was spitting bloody, frothy sputum. She laid her head on Pam's shoulder and died. She waited a little too long to treat the congestive heart failure. All of the respiratory support, medicine and shocking does nothing when the Lord calls you home.

The ER waiting room looks like a war zone. There are extra chairs up and down the hallways. A young girl in a wheelchair has an ice pack on her knee and tears streaming down her face. There is a small blonde child with a bandage around his forehead. It may be a small wound but it needs repair. There are two young girls crying because their Mother is going to surgery—to have her appendix out. The young man is slumped over the chairs with an emesis basin held in his hand and retching. There are pale, anxious and crying babies and restless young children.. The charts of the patients are backed up—10 to 12 deep and everyone waiting anxiously for their turn. We need more ER's, more Doctors and more nurses. People waiting, waiting for their turn! The 73-year old lady must be filled with fluid. Her feet and legs and abdomen are distended—tight. There is a 15-year old boy with diabetes—"All my life." And he looks thin, hardened, alone. He has a blood sugar of 687 and epigastric pain. Some of these patients wait because the ER is full of heart patients—seven in a row now. Is it the "end times"? When men's hearts are failing them? The 63-year old male has a history of collapse of the mitral valve and his blood pressure is only 49 on a dopamine drip. He is cyanotic, restless and vomiting. He is complaining of pain between the shoulder blades and is cold and clammy. He needs a balloon pump and transfer to the cath lab. He is just not going to make it! The 55-year old in bed 5 has had "bad indigestion" for two nights. He has an inferior wall MI with rales in both bases of the lungs. He is already in heart failure. He has pain that is an 8 on a 1-10 scale and the morphine has been given and the Dr. has ordered to increase the nitro drip. He is vomiting, even after the Phenergan has been given. His heart rate is 126 (normal 60-80) and we just can't get his pain relieved. The TPA—the clot buster drug is being missed to give 15 mg bolus over 5-minutes, 35 mg over 30-minutes and 50 mg over one hour. He is pale, ashen, diaphoretic and restless. It is hard to watch the face of suffering.

The 44-year old male in room 2, "He unloaded something off the truck yesterday and started having a slight headache. He awoke with loud snoring and a sever headache and was unable to stand or walk so I put a cold cloth on his head." He died with a clean face!

The patient states, "I have a coffee crotch." She has hot coffee burns with edema of the labia and peeling skin. The Dr. has ordered a shave, debridement, Neosporin and a catheter because of the swelling—OUCH!

The little 85-year old lady has arm and shoulder pain and is worried about her heart. While checking for pedal pulses for circulation and edema she states, "Honey, my feet don't hurt. " Could it be the calcium injections she has had for couple of weeks?

The little 4-year old girl fell at the pool straddling a step and sustained a small perianal laceration approximately 1 and 1/2 cm. The child is calm and cooperative and the Mother is crying and wringing her hands. "Is she ruined? Is she ruined?" No, but she will be Mom if you don't straighten up.

The 59-year old in bed 2 has chest pain and is very talkative and anxious. He has a coronary artery bypass six years ago and then got a staph infection. He had the sternum and some muscle removed. You can see his heart beating through that tin layer of skin over it. The deep muscle tissue was removed and he had skin grafts over the chest. He wants to know, "What that old heart is doing?"

The diabetic in room 7 is only 40-years old and he continues to get chopped up. First the legs were amputated to the knees and the fingers—two are rotten now and he had had 3 strokes, 2 heart attacks and renal failure and he doesn't complain. Do you wonder about his diabetic control?

There are 10 beds in the critical care unit and the oldest patient there is 57. There are two 20-year olds, one with an acute appendectomy and the other with acute pulmonary edema. Look at the young beautiful girl with lily white hands. She strangled herself with a belt and is in a vegetative state. Will they pull the plug? There is a 20-year old with sepsis and a lung abscess. He had a pneumothorax from a puncture wound to the back with a nail gun.

In room 4 is a little boy with a little broken arm. He had a little fall from a big monkey bar. He will be fine and maybe his Mother will survive.

And next are two little boys who were going down a slide at a water park. Both have little parts caught in the mesh of their swimsuits. The thread entrapped them with the worm floated through.

This had been a sad baby day: a frail little 5-year old in bed 3 has soft tissue sarcoma. She had a feeding tube and is vomiting. Her abdomen is rigid. She is dying.

The 9-year old has psychosis. He is on Mellaril, Lithium and Ritalin—some old and powerful drugs. He was chasing his cousin with a knife. He said he hears voices that say: "kill people, kill people".

The 23-month old has popcorn up the nare. The nose is swollen and we have attempted three times at removal with suction, blowing and alligator forceps.

The 6-year old is 2-days post op from a tonsillectomy and he won't eat, drink or take the pain medicine.

A young construction worker fell into a 4-inch pipe. He as a gaping wound in the chin plus he is pale and sweaty with a low blood pressure.

A new day in the ER and it starts out slow—did I say that? Two females come in, overdosed. One did GHB and never responded to all our attempts to help her—so young, so beautiful. The other will be fine if she gets away from those drugs.

A day of really sick people. One gentleman has an aneurysm and was sent to neurosurgery at the big hospital by helicopter. The last patient was a 50-year old and weighed 365 pounds. He was pale, sweaty and yelling with abdominal pain. He was incontinent and his blood pressure kept dropping. He was bleeding out from? The Doctors work hard to put the puzzle pieces together and arrive at the right diagnosis and treatment.

The next patient was a homeless man having an MI. He was dirty, pitiful but very grateful. He was transferred to the heart catheterization lab.

An ER day full of code blues—three of them today and two of the patients died. On was a 45-year old construction worker who had a crush injury from equipment at his job site. One was a 70-year old man suffering with lymphoma and the third had a complete heart block and was kept alive by a pacemaker.

I was called in to work at 3 AM and that makes a long day. There are three young kids just brought in to the ER (ages 14, 15 and 16). The have used acid, GHB and a lot of alcohol. So much for Halloween Horror Nights! A real horror for their parents.

How can 8 people all get chest pain at the same time and the ER is already in overload holding patients for admissions to the floor?

The young man presenting at triage has a list of complaints He has, "generalized body aches"—that is the complaint of the day but he states, "My quadriceps and biceps are sore and I have pain from the solar plexus to the clavicle and my hair follicles are sore." He must read more medical books than I do.

I have a family here looking for Grandpa. He was sent in by an ambulance, "because he wouldn't eat his soup." There is soup gurgling in the back of his throat but Grandpa is very dead.

The ER sees a lot of people with back pain and lot of those have the DWS—I don't want to work syndrome. They want pain meds and a note to be off work for about six weeks.

Some days some people make my heart hurt. The freckled 23-year old male is so thin and all his fingertips are rotting off. He has a history of Raynaud's Disease and Scleroderma. The fibrous tissue turns to scar tissue instead of normal tissue.

It is 7 AM and the day shift is greeted with a 26-year old body in bed 2. There was a trauma code from a massive head injury. He is tall, dark and handsome, young and dead. His buddy is in x-ray for a CT scan of the head. He has a head injury with facial lacerations and was the drunk driver of the car. He comes from x-ray complaining because his shirt is messed up. The police want a statement before we tell him that his friend has died.

The morning proceeds with the frightened, petite blond female who celebrated her birthday by using cocaine and ending up with an overdose.

And the 28-year old unresponsive black female—just drunk?

The radio blares with good news. We have 12 patients coming from a car wreck and we already have 5 patients waiting to be seen. Please—we can take 6 but send 6 to another Facility. Hey team, I hope you had a good breakfast because I don't think we will have breaks or lunch today.

There are three member of a family from one motor vehicle accident. Where is the Mother? She slipped off the backboard and went for a cigarette—with neck pain and tingling in her fingers. She is willing to risk paralysis for a cigarette. The Father states, "Here, take the baby, I need to check my equipment in the car." Priorities—pitch the baby on the bed somewhere—I need my stuff.

I ask the same question doing a history—"Do you have any medical problems?" "No, just ugly." Should I agree with him?

That man is from a MVC and is not speaking. He is one very

large man and looks very drunk and it is 10 AM. Shall we make the morning interesting and take bets on his blood alcohol level?

Chapter 5

Personnel Pranks and Private Thoughts

This E.D. needs a morale officer. There are crazy shifts and nurses are being replaced with techs; there is not enough help, no raises and the management is worried about competition with other hospitals. It used to be a fun place—at times anyway. Nancy Nurse was a lot younger then and not such a seasoned veteran. All the good nurses are leaving and finding other jobs. Am I going to be the only one left? Hey there, you guys in the plush offices at the top of the tower. Look! Look! Can't you see the quality of care going down the drain? Don't you even care? I thought patient care was a priority—not computers, numbers and dollars.

It's the flu season. There are extra chairs lined up in the halls, the waiting room and an occasional sick on lying on the floor. The sick you will always have with you—looking, coughing and waiting. The hours slip quietly by and one by one the roll in and out.

The 75-year old female with a bright copper hair dye job has been on slim fast for two weeks. She now has a fecal impactions. She's trying to shape up for a new boyfriend. Where else can you get the old soap suds enema—the three h it is called—hot, high and a hell of a lot. Oh, please lady, go back to eating a normal diet that keeps things moving.

Managed Health Care—-#$%^&*! This patient has been in the ER since 5 AM and it's now noon and her insurance carrier wants her transferred to another facility that has no critical care beds available. So my patient with acute chest pain lies on the ole gurney in the midst of noise, illness and confusion to await further care. "I'm her

Doctor and I want her transferred." Would you help us and order up a bed and a new room. Sure, we can hold her in the ER that long.

The Doctor today is very slow—having a bad hair day or just doesn't want to be here. Could you just have your heart attack another day? The nurses are fighting and it's a holiday weekend. There is strung out Sam and bulldog Pat; he's had to many uppers and she's angry at the world because she's been replaced by a young lover. The supplies haven't been stocked and the paramedics have carried off all the pillows or maybe the pillow monster struck again. We have been issued 800 pillows this month and every morning there are half a dozen missing. Maybe we can tag them like the department stores do the clothing and a loud noise will sound when they leave. It is so discouraging not to be able to find a pillow just for patient comfort.

Blood is dripping down the bed rails and housekeeping thinks the nurses should clean them because this is a holiday weekend and they are short staffed. The blood can dry and hang around until Monday and hope that the last patient didn't have Aids.

We did bring in holiday goodies but every time we try to have a little party in the lounge a bunch of sick people come in-duh. Nursing is such a wonderful profession—where else can you work 365 days a year and 24 hours a day!

Yes—we were reported to management for talking of non-nursing things at the nurses station. Humor sometimes is our only salvation to bear up under so much sickness and tragedy. If only that patient could have heard the whispers at the Doctors desk. Dr. Jays asked the physician assistant how to spell epididymitis—the answer—nut. We has another patient with belly pain: wait, I have to eat lunch first. The little redhead, our Nurse Manager would jump on your tombstone if she heard some of these things.

We can't write on the chart what we really think!

1. He's strange

2. 2 screws loose

3. Nice buns—how old is he, is he married

4. She's strung out—way out

5. Drug seeker—expect a drama code

6. He's a drunk but only had 2 beers—right!

7. Grouchy old man—how does his wife do it

8. B.O. big time—have those socks never been washed

It's 4 AM and Dr. Frank is called from his bed in the lounge and I quote, "Tell them to bleed, suffer, die or whatever, I'm asleep." Cath the patient and send a urinalysis. I know that procedure gives you 45-minutes more to sleep but this patient only has a sore throat. Well, they used to take a nap on the extra bed in the trauma room, until visions of the mist that they see when they close their eyes. Dr. Amstead said he would never close his eyes in that room again. Ghosts of Traumas past.

Hello, is this the Emergency Room—my child is really sick with a high fever and cough but I know he's teething. Yes, please bring him in we will be glad to see him. "But I don't have a car or any money." Nancy wonders what kind of miracles we are supposed to perform over the phone. And every child's ailment in the world is blamed on teething, from 105° fevers to rashes and abdominal pain.

And then some of our patients are such frequent fliers we know them personally, their families, how they like their pillows plumped and how much sugar they take in their coffee. A middle aged (looking) regular named Nell is on the phone stating that Garrison—her sweet 80-year old husband is chasing her around the bedroom. Maybe he'll catch her and she won't have to visit us today. She really carried in a dead chicken by the neck and told us it was sick. It looked sick—didn't even have feathers. You wouldn't believe how man telephone calls we get each week looking for missing persons. "Have you seen my husband, lover son or neighbor? I'm checking all the hospitals in town—he didn't come home last night. Sure, I'm only checking in six people here in triage at the moment and one is dripping blood on my desk and the guy in the wheelchair is sweating profusely and looks like he may be having the big one.

A wounded patient arrives and we usually get 49 repeated phone calls and fourteen people sitting in the waiting room who want to watch their suffering. But, I would be there for my loved one too.

Phones are ringing—never ending. Oh those telephone calls. "I'm going to kill myself." Nancy gently tries to keep him on the line while writing a note to co-worker Patty to have the call traced. It never works to transfer them to the psych liaison person—the call gets lost and you lie awake at night wondering? Count sheep—no count narcotics—that monotonous thing you do three times a day to keep track of drugs—in case your fellow nurses are stressed, can't cope or have chronic head aches or back problems. Well, that's the kind of things that drug seekers tell us.

Some days it seems everybody wants drugs. I need a scrip for Valium. I have been on it for 14-years now. I have had headaches since age 12 and only Demerol works. The wonder drug Imitrex never for works for the Demerol crowd.

The Emergency Department is an exciting place: drama, first hand news, sudden death, code 90's, handsome doctors, tired feet, aching backs, too many sick people and not enough help. Terrible tragedies and exposure to every ailment known to man. It is a definite calling and not just a job.

Poor Patsy Hamilton's drug addiction is a way of life—slurred speech and a grating voice. "Remember, I weigh 200 pounds and I need more medicine now." And she even sends in her nephew requesting the same drugs. Repeat offenders, how do you know who's lying or true—you can't see their pain.

My friend Mary said she didn't sleep much last night as she was volunteering with her local rescue unit who responded to a motorcycle verses tractor trailer accident. She picked up a helmet at the scene and the head was still in it. One decapitated teenager. I didn't sleep much either the night after the young redhead came in from a motorcycle accident with the bones of his right leg extended in the air with no meat on them and the injury extended into the rectum. He clawed my arms as we started the IV lines trying to replace the blood draining away from his body and screaming as the Doctor tried valiantly to clamp the bleeders to give enough time to get him to the operating room to amputate the leg and make a colostomy. He lived to ride he motorcycle with one leg. Or the night the 16-year old boys who had watched a TV show and tried to play Russian Roulette. Only one handsome young man had half his brains blown away. Or the night after helping the mother identify her 10-year old son by his jeans and sneakers after his bicycle had been struck by a trailer truck. He proceeded finally to brain injury rehab—but he will never be the child she knew before.

The only time I ever felt really ill during an emergency I thought I would die—nausea and I felt faint but I took care of the patient first. My little sister came in to the ER with her now amputated. She was helping her husband and my Dad to install an attic stairway when it slipped. My Mom was home crawling on the floor looking for her nose but it was still attached with a little tag above the lip and a large laceration below it. My sister's nose healed after plastic surgery that night. It is a little perkier but the memory of that incident remains. Deana's Father came in Code 90—he died. Susie's Father came in with amputated fingers from a skill saw incident. Yes, it's hard deal with emergencies with those we know and love. When the radio sounds and the age and sex is the same as your teenager or spouse you can't help but wonder.

The Med Com Radio announces a new patient—suffering with anxiety and it's 6:45 am. It's torment to be anxious because the mind plays tricks on us and it is so hard to determine and treat successfully. But this time the patient used cocaine last night and his heart is racing. How could our favorite medical examiner look at those bodies and diagnose death y crack cocaine and go smoke crack. He was a Code 90 in our ER and our best efforts couldn't help him; the racing heart stopped, never to go again and M.E. died.

The patient in bed 6 is not paying his bill because we didn't notify his wife that he was in an auto accident. Facts: He was very stable and had a scratch on his elbow, he has to walk home, it doesn't matter that his ambulance was met at the door, his vital signs immediately taken and was undressed and had a full assessment.

Was seen immediately by the Doctor and had x-rays ordered and taken. But our care was sloppy and unprofessional per his accusations. The accident was his fault and he got a ticket and three other cars were involved. I guess we are the scapegoats.

No, No, I can't face Mrs. Gentle this early in the morning. She is a frequent flier and always needs the bedpan on arrival and states that her arms are too short to wipe herself. We almost need a crane to get those cheeks elevated enough to slide the pan under and her sister and daughter accompany her and so sweet and demanding. I don't know how she goes potty at home! And I will never understand why people can't talk when they are sick. It's hard to fill in the history with moans. What will the diagnosis be today? Chest pain, constipation, a freckle on her knee? Maybe, I'm glad she's not talking today to remind us that we have to do what she wants because she is paying the bill. She is a rare bird and sometimes we just need to vent.

Short of breath, skinny, wrinkled, Mrs. Peaches just wants us to admit her and let her stay in the hospital until they find out what her problem is. Couldn't be those 2 1/2 packs of cigarettes she smokes daily or the fiery liquid she smells of causing her problems. *Choices we make—consequences we suffer. So much of our health issues we do to ourselves with lifestyle choices and immorality.* The health care industry has made great advances and can do many wonders but we live in the fast lane with a diet of fast food, fat grams, couch potatoes with the remote changes in our hands along with the car keys to run out for more highly processed sweet, fatty food and many disease are on the rise due to obesity and sedentary lifestyles. Carrying an extra 100 pounds and puffing that lovely smoke fro a sedentary lifestyle. What does it take to reach that TEACHABLE MOMENT! Very often it's only after the stroke or heart attack.

Did you ever wonder about those multiple sticks for blood? Ok, it's a secret but there are good reasons:

1.   The nurse put the blood for both cultures in the same bottle instead of 1 cc in each bottle. Whoops.

2.   Hemolyzed blood—the labs excuse for whatever happened to the specimen.

3.   The blood was put into the wrong tube and the purple top with diluent was needed for the CBC.

4.   Not enough blood was obtained—poor stick or did you five already at the office?

5.   Somebody forgot to attach the label and the lab won't accept the specimen.

6.   The blood was labeled with the wrong patients name.

7.   The specimen was dropped and broken en-route to the lab.

8.   The doctor added orders one at a time.

Just remember that your health care givers are just people and we all make mistakes. We just pray that ours won't be life threatening. And you know it takes just 10 seconds to tick off the patient and they remember it for ten years and 20 of their friends and neighbors what a terrible thing you did.

Nurse, "I need water!" "I'm sorry, we need to find out what is causing the abdominal pain first." If she went to surgery and died of aspiration she would really be angry. Nurse, plump my pillow, another blanket, put the head of my bed up, when is lunch being served, is my room ready upstairs yet, call home and give my family the room number. All this as the other patients pour in the door, lie on the floor and hang onto the walls wondering what those nurses are doing in the back. I do agree with one thing, it's cold enough to hang meat in here, thanks to those middle age, hot flashing co-workers who continually flip the thermostat to 50. Maybe we should strip them.

The sick you will always have with you! We need six more nurses, 12 more stretchers to put patients on, 3 more Doctors and a lot more patience or fewer patients. I don't know why people can't schedule their emergencies a little better, say one q-15-minutes. Patient Sandra Dooley is on the phone and complaining about her treatment from six months ago. "It's now July and she can't get her bill straight either—the insurance is not paying."

Were is Dr. Davies—he has been in Room 7 three times already to check out those boobs: I mean injured ribs. Double D, tangerine t-shirt—he's taking the P.A. for a second opinion. And a certain Mrs. Hooper ahs called every day this weeks to see is Dr. Davies was on duty. She always need advice on some medical problem. But every young blonde patient gets to see Dr. D. first, don't send in the P.A.—seniority. And life goes on and on in your local Emergency Room.

There is no room for boredom in this job. Sick people all look different and have a different story to tell. I didn't know that today I would take a wild ambulance ride through heavy traffic with a 22-year old pregnant female with imminent delivery. She has had no prenatal care! Lights, sirens—baby?

The med com announces another: Gunshot wounds with CPR in progress. In be 2 we already have a 12-year old with near drowning. They are on vacation and he can't swim but took a dip in the motel pool.

The patient in bed 3 is having sever chest pain and needs the cardiac workup and treatment.

The 27-year old with the gunshot wounds was in a lovers spat. He must have been a little ticked—there are 7 entry wounds to the chest—we are inserting the chest tube on the right as there is gross blood. We need the heart tray. "Let's crack a chest—where are the rib spreaders?" There is a lot of adipose tissue, but the lung exposed is nice and pink. There is a hole in the heart—with the life's blood flowing out. No go—she's dead.

This has been a full moon kind of day. From the depths you could never accomplish a rerun of this chaos and madness; of the sickness and death; of the pain, anger, grief and shock of sudden tragedies.

In bed 4 is a 37-year old Aids patient. He overdosed on Valium—he said he was upset with his roommate—for giving him Aids. He is restrained to the stretcher with four point leathers and covered in charcoal, he was supposed to drink to absorb the medicine from his system. He obviously drank some of it as there is black diarrhea which squired the bed at frequent intervals throughout the morning as we waited for a monitored bed for him.

The next patient, Kelly is almost 5 feet tall and weighs 265 pounds. I wonder why her knees hurt? She has a pin in the left breast—that's a lot to hurt. She asked if I would lift her onto the stretcher—were is that stool?

We have a chest pain sent from the clink with pneumonia. The onset was Saturday with nausea, diaphoresis and tightness in the chest. That's not pneumonia—a myocardial infarction and it's too late to give the clot buster drugs.

The young black girl in room 8 needs a pelvic exam per Dr. Dooley—"there is a grossly swollen twat." Per the patient "a risen on her monkey". Actually we have a bartholin gland cyst which is a large abscess which need to be lanced and drained. Do we have to— before lunch?

Well, hello Miriam—I haven't seen you for a month or two. She had the gallbladder out and we never found a cause for the chest pain, but finally the depression is being addressed and treated. She is on Zoloft. The source has been found for this multitude of vague and disturbing ailments.

The 45-year old male in bed 2 took an overdose even if just a little one—a little cry for help that Lopressor and Lorcet won't fix. He was fired from his job for sexual harassment and now his wife may leave. "I just can't cope." I am so sorry we have to put down the Ewald tube—it does look like a garden hose which is pushed in the mouth and into the stomach. The Dr. just said he would break his teeth if necessary to get it in. The patient wants the hospital administration and rightly so. He has emotional problems and is now being abused by the M.D.—he is already a threat to his own life.

The young man has a clavicle fracture from a wrong turn on his bicycle. Dr. Dooley put the clavicle strap on upside down and the loops through the wrong hooks. Now let me help you adjust these straps.

The physician assistant is here working this morning is sick himself. You macho men need to stay at home when you are sick and give the bug to all your nurses and patients. He is complaining of waves of nausea and can't really listen to the patients tale of we. "Excuse me"—a rolling tummy—clear the restroom!

She's 43, fat, neurotic and frustrated. You can tell by her looks: list of allergies—to Haldol, a major antipsychotic medicine, Prozac, blood pressure medicines and sleepers. He has a split on the right hand and said the fingers are always tender and nobody better touch that hand. "I bumped my head four days ago on a car door and I am getting headaches and they are ruining my vacation. I had to wire my husband for more money and I need drugs. The Dr. said he wished his Rx would say, "Get a Life!". "Find something worthwhile to do." When we look back on our life what will we and others remember?

Hyper Stan—You're wiggin! Did you have too many pills today? Someone's stash fell from the light in the break room bathroom. Who were those pills intended for?

The operating room is ready for the patient with a rectal abscess. You have three minutes to start an IV, draw labs, start antibiotics, get an OP permit signed and get and EKG and chest x-ray done. Please admit the patient, which involve explaining and signing a living will and he is still fully dressed. His nurse is feeling SOB— short of breath.

Is this pediatric day? Things do seem to come in threes or sixes or sometime by the dozen. We have 6 crying kids, 5 broken bones, 3 sore throats, 2 MI's and one hypochondriac. Please put in a call to Dr. Joys to come in early. "I'm dripping we and just getting out of the shower." OK—you can dress first.

I need a potty chair for the gentleman in bed 4. Oh no, there is a very old specimen brewing in that chair stored in the dirty hopper room. Somebody is passing the buck!

I love my job, I want to grow with the company. I want to be a team player! So, I go on to carefully remove the IV line on the sexy blonde with Pyelonephritis to send her on her way home and then she vomits all over. No discharge in bed 8. Restart an IV line.

Little Mr. Williams is ready to go home and he has no ride. His housekeeper dropped him off and said, "don't call me, I will call back." I don't think he is able to manage a cab and would his housekeeper let him in? There are no other number on his chart and he doesn't have a clue.

There is an unfortunate delivery—for us—a 400 pound back pain. After turning him for the injection I now have back pain. Is back pain contagious? Sorry, he does have a name and a face and he is in terrible pain.

I'm glad the oriental male in bed 3 with measles didn't understand me when I picked up the chart of another patient and tried to ask about his rectal fistula. I told him to undress and lay on his left side—to check a rash?

My patients have some interesting home remedies—take or eat garlic or garlic pills every day to stay healthy and treat high blood pressure. Well. Mrs. Toothman is slender and active and mentally sharp and 92-years old. I can't say much about her breath. The baby has raw potatoes on and string around it's neck to help teething. I hope the string doesn't get tangled.

Well, young parents and I guess even older parents. I have one old nurse remedy. The best thing you can do for your children is to love your mate. Fathers are not indispensable and the arguments, separations and divorces wreck havoc on these little people.

My child has pig eye—could you mean pink eye?

The dark skinned male in bed 5 has a blood sugar of 427 which is about 4 times the normal and has his medicines as heroin, Percodan and cocaine. "My Father has lots of money." He is so combative we need the application of leather restraints. He states, "You will be sorry!" I hope the emergency buzzer still works in the triage areas as many threats as we have had lately.

The MD looked at the patient as we initiated a Code 90— "You're going to be ok." Under his breath his me he said, "He is on the way out, he's going to die and going down the tubes fast." We couldn't help him!

The Jehovah Witness patient has been bleeding vaginally for three weeks. Her hemoglobin is drastically low--only 5.1. She is a few pints low. Her husband said only comfort measures. Don't give her blood. Let him look death in the face. She is too young to die.

The next patient we beat in the face with a pipe. He is so rude, I know why.

The ears are plugged and he can't hear out of the right ear. We will ear-i-gate. Actually we will irrigate the ears—wash the wax out and you will be good as new. "Did that actually come out of my ear?"

Do you know this is the sixth patient this month who claims to have lost their Rolex while in the ER? Now a set of teeth, ok—we do remove those for various reasons.

"I have to take my clothes off?" "It's really hard to examine you and do a pelvic exam when you are fully dressed." "A pelvic exam—I don't need it, don't want it and can't pay for it." We can't change the oil without opening hood.

Where are the labs? It is done in the computer—but was it done to the patient? "Move the Meat"—did the MD really say that? Look out there—there are sweet babies, really nice people, accidents of all kinds, lacerated hands, broken ankles, people from the motor vehicle accidents and a little gentleman that said, "Gee, I didn't clean up before I came." Please don't take time for a shower if you are having stroke symptoms. Falling in the shower and pulling the soap dish off the wall causing multiple lacerations to the foot only compounds the problem.

She is blond, beautiful and drunk. Maybe was beautiful. Her hair is stringy and she has puffy red eyes and can't hold her head up. Her husband states that she has been drinking for five days. ***The vices that get you—too much of a good thing—takes you further than you wanted to go and keeps you there longer than you wanted to stay.***

Just for human interest—isn't people watching fun. 58-year old Patricia has male pattern baldness. She has hair just at the ears down and has it flipped under with pearly dangle earrings. I be she is all of 80 pounds. I love to guess the age and weight of the patients as they check in. I usually can tell if they smoke just by the smell of the gray pallor of the skin or lots of wrinkles. Ladies—do you know that smoking ages your skin 10-years. Even the miracle creams can't conceal it.

Is the illiteracy? A 29-year old man from Kentucky has slashed wrists. "I did it on a razor blade—you want one?" "Do I want a razor blade?" "I have them in my back pocket and more in the front. I can't read or write. I had a bad childhood. Do I take pills? Sure, I take them but they don't do any good. Do I use recreational drugs—I did crack at 5:30 this morning, but none since. Do I drink alcohol—usually no more than 6 beers as day." Now how do I retrieve the extra razor blades so that he doesn't finish the job?

Let's set up for the pelvic exam. The patient has a "tingling above the clitoris", don't show your surprise. It is tinted and shaved—is that a reverse Mohawk?

It's a smelly day all the way around. She's 40—ashen gray in color with big pitting dark lesions on her body—muscular dystrophy? She can't keep her leg on the bed and she is suffering with vomiting and diarrhea. Gross—I wish she could use the call light for a basin or bedpan.

In the next cubicle is a little frail 63-year old lady. I tried so hard to be kind and nice. She has a warm blanket, call light and Gatorade to drink. She is screaming—blanket, blanket, blanket—I am freezing. She has a gastro bug and is so upset to be discharged home. She said, "My family is out for the evening." She does need care, fluids and a bathroom close by, but we have 20 people waiting for an exam room.

That five year old boy is nursing—standing up to his Mother's breast. Extra chairs, extra patients, lying, sitting, bleeding, moaning and vomiting. Smelly, short of breath and crying—the sick you will always have with you. Some days going out of those ER exit doors is like going out of the darkness into the light.

A grown lady—a 36-year old mother has a simple knee laceration. She became hysterical when I asked Dad and the six-year old to wait in the waiting room while we repaired the wound. She is afraid the child would touch something—contract something. I hope nobody passes out on the floor during the procedure.

There are 20 patients in treatment rooms with 12 waiting. No matter how many patients we see today there seems to be always a dozen waiting. We have called in extra nurses and another Doc. It's 80 degrees outside with beautiful sunshine. How can so many people be so sick. I know, sometimes bad things happen to good people. And accidents do happen—often.

Look at the pale little 17-year old with a peritonsillar abscess. We lanced the abscess and she is spitting purulent goobers. Start IV Ancef and Solumetrol. She is one sick cookie.

Can't her husband speak or won't she give him a chance? The little thing—the wife. "It's our 50th anniversary and the kids are at home. See him now and we don't want to spend our whole day here. He don't need an x-ray and what is that medicine? You mean we have to go to the pharmacy?" This is what my old Doc friend would call a horsey woman.

The young male has an ugly elbow wound with a tourniquet made from a bandanna. We did remove the tourniquet and did a pressure dressing and the bleeding is contained He is pacing the floor, yelling, "Send someone for me a cup of coffee." My arm is bad, call engineering my arm is bad. Sir, is that for coffee or to repair the wound. Later as he see the morgue cart roll by—"Who was that?" "The young man we saw ahead of you."

Patient complaint—new onset of sexual activity. Now she has a new onset of pregnancy, new onset of herpetic lesions and new onset of vaginitis.

It's 7 AM and that's a weird looking guy in bed 7. He has

shaved eyebrows, twigs of fuzzy long hair, smooth skinny legs and his nipple bitten about off.

This is just how I wanted to start the New Year—called in to work at 3 AM. This is only the busiest weekend of the whole year. The local MD offices are closed and the ER running over. Holidays add a lot of strain to everybody's already busy schedule and this seems to be the hottest celebration spot in town. We have blood, excitement, confusion, police checking out the motor vehicle accidents and flu bugs from all over the world. Drug seeker mixed in with pinworms and lots of belly pains. A typical day full of sick people. Some nice, some rude, some weird, some old, some dying, some smelly, some little and sweet and some old and gruff. Some henpecked and some lonely. I really didn't think about spending all my holidays at work.

I love coming in to the ER in the early AM—to the sights and smells of home. The patient in bed four is coughing, sputtering, vomiting and gagging—and not quietly. The alcoholic in bed 3 is 3 1/2 times legally drunk. "I'm ok. I've been drunker than this before."

And then we have the stream of Webster County fire fighters going to visit their Lieutenant who came in as a Code 90 five days ago and is still unresponsive. Some we wonder—saved for what—a vegetative state and massive hospital bills. They tried so hard to resuscitate him.

Now back to the Code Yellow—search all areas and see if things look normal. Her some the SO (Sheriff Officers). Put all the ambulances on stand-by and security at the doors. Will we evacuate the building? The dog brought in with the sheriff is sniffing at the paramedic's office—which nurse is on the outs with her boyfriend?

Just keep doing what you are doing—suture the little laceration. You may be blown to bits in the meantime but we don't want to panic anyone.

Zap, Zap, Zap. Big Zap, little Zap, everybody Zap. A mnemonic to remember the protocol for cardiac arrest resuscitation. Save heart muscle, O2, 3 IV lines, EKG, Chest x-tray, draw labs and give aspirin and start a Nitro drip. Did you have to tell me this stuff costs $4,300.00 a bottle? What if the bottle slips? What is the money—compared to a life and a future?

Mothers, please keep your newborns at home. Keep them well. If a baby comes in the ER with any fever two months and under we have carry out the protocol. Blood sticks, urinalysis and a lumbar puncture. I have to hear that 35-day old baby cry while I hold her for the procedure.

The screams are loud and long, ringing throughout the hallway. A bowel regimen, please for this downs child who is constipated.

Infections, open wounds, HIV, TB—Wash your hands, wash your hands, wash your hands. How did I become so obsessive compulsive?

The view from the triage desk isn't so pleasant this morning. There is a big, loud man with his belly hanging over dirty jeans. He is sporting a droopy mustache and has two little unkempt children following along. "Where's my wife?" Well, sir—"What's your wife's name?" "Louanne MacDooly—they sent her here." "Let me check sir."—through the computer file and a room search. Sir, we have no one by that name. "I'll just run over to her job." To the children—loudly—"Sit down and shut up." "Would you like to call?" "I don't have that number in my head." "Could we look in the phone book or call information?" Angry and anxious doesn't always think clearly. Please don't leave the children. There are several hospitals in the area or maybe she is en-route.

Emergency Rooms—a clinic for welfare patients; out patient treatments or geriatric care. Emergency care touches on everything and everyone. Constipated or Conjunctivitis—old and young—from birth to the grave. ER's see all sorts of trauma and mental and physical ailments alike. A new found breast lump or the peace of relief from an asthma attack. The sick you will have with you!

To the ER personnel sickness and trauma become almost routine but some of those patients live forever in your heart and memory. Nothing exiting her today. A screw in the leg—"he screwed himself." Do we need a screw driver or something more surgical? A chopped off finger—very traumatic for the patient but we put the fingertip in gauze over ice and call for a hand surgeon. Oh yes, there is a black lady screaming in the doorway. There is an 11-year old in her car, unresponsive! The child woke up and then is out again. The Mother is screaming so loud that all the other patients are peeping out of the curtain cubicles. There are heart monitors beeping, phones ringing and all the Docs and nurses running to the patient.

The morning here in the ER was nice, as few simple ailments and a broken bone or two but by afternoon there are disasters left and right. WE had one flown out by chopper to a larger facility and the cath lab and probably surgery: a myocardial infarction with a complete heart block. The next patient is having a stroke and has no speech and no swallowing reflex.

There is a 30-year old—Dead? His wife state, "I tried to wake him and he was dead. I gave CPR. Why did he die?" The drug panel shows cocaine and pot but he is now alert and has stable vital signs. Will he learn from this near death experience?

We have a psych patient in room 5 with a Baker Act—involuntary admission. He was found huddled in the corner of the treatment room with 2 scalpel blades. Thank goodness he was found in time—back in the restraints please.

A tall black person presented in triage with sinus congestion, earache and dizzy—also sporting boobs, tall heels, an uplifted hairdo, slim dress and long nails. I asked as per routine, when was your last menstrual cycle and it said, "I'm a guy."—oh.

A day of tribulation—the masses—sick babies—strange visiting Doctors barking out strange orders and harassment from a lady with a bus load of students and two who needed treatment. She pushed and shoved until I told her to leave me alone—nurse abuse.

There is yelling and screaming from a psych guy with a drug addiction. He has huge knots on both hips from scar tissue—too many injections of something. He threatened to kill us all. The MI guy is having more chest pain and yelling for an alka seltzer. I love my job—I want to grow with the company. I want to be a team player—most of the time.

Another near drowning—age 2. Will this child have permanent brain impairment? What a difference a few minutes make.

Choice in life have costs—consequences. There is an untimely death—fatal choices from bad habits and some are only wasted lives ending in destruction.

That gentleman was wandering at a local motel. He has a passport from Spain and stated that an alien was implanted in his Mother's womb at 3-months and the he still has contact with aliens.

A very busy ER early—a code 90—he was having an MI and had a sudden V-tach with in not compatible with life for very long. It seems like I couldn't stick anyone and needle two or three IV attempts. Most of my coworkers are hung over from Dr. Remney's party last night. A new traveling nurse asked for another assignment—after only three days. She said we need more space, more help, more Doctors and better equipment.

I am too tired to wiggle much less keep my appointment after work.

The 36-year old in bed 7 had a hemoglobin of 3—several pints low—and a perforated uterus. She had a molar pregnancy 3-weeks ago and leaned over the tub to get her daughter out and felt a sharp pain and passed out. He phone was out of order and she used a garage door opener and drug herself outside and a neighbor who was out smoking hailed a passing car with a cell phone. I just wanted to get her to the OR where the bleeding could be stopped. Those kind of emergencies make me sweat.

I was greeted this morning by two psych patients in leather restraints. One is hissing at me and one said, "I am Jesus". Drug psychosis is not connected to normal thought processes.

The 35-year old male said he had his girlfriend "circumcised me" yesterday. He is in pain and bleeding. "They just didn't do the suture repair." states the urologist. Call for a psych consult.

Two young women come in at the same time with drug

overdoses. The first is a beautiful 18-year old blonde with a navel ring and purple undies. Her Father called to ask if she was dead and her Mother is tearfully waiting the outcome. Her boyfriend was arrested yesterday with cocaine possession. The friends of girl number 2 come in to triage saying "We have a friend in the car who doesn't look good. She is turning blue." She is a 21 –year old with no breathing and no pulse. Bring a stretcher—get the code cart! Will she live? Will she be brain dead? Her friends state that she was staggering when she came in last night and wouldn't wake up today. As a Mother, my heart aches for the girl and her parents.

The ambulance puts a patient in bed 2 who is pregnant with an imminent delivery. Dr. Avery states, "We don't do deliveries here." (No OB wing at this hospital) As we bring the OB tray and the baby isolette to the bedside. "We don't do deliveries here."—as Nancy unwraps a gown for him. Put this on and she hands him gloves. "We don't do delivers here." as the paramedic catches the baby. It's a boy. Now he will deliver the placenta and take care of things.

The next patient presents with rectal bleeding. "I have this piece of meat hanging there." Is the bowel telescoped or is it a hemorrhoid? We will take car of you.

I don't know how some children survive those parents. The yell continuously or are indulged with pacifiers at age 4 or a bottle of soda. The child has had "a fever of 105" for three days and has had no treatment. I think the ultimate abuse is to set no limits and have no restrictions. And then there is the Mother who treated her critically ill son with so much aspirin and Tylenol that he had a bleed and nearly died. Often the ones we need to see don't come soon enough and the sniffly, runny nose playful child comes in twice a week.

The 43-year old male in bed 5 said he didn't come in earlier because he was on vacation—even though "It's the worse headache of my life." An intracranial hemorrhage is diagnosed as he stops breathing—respiratory arrest—Code 90—never to start again. Too late, too late—a vacation that will never end!

I know it's like a turtle—this dead body is just a shell. The little angel who used to live here flew away home—to the waiting arms of Jesus. My tears are difficult to control as we put the little body in a plastic sheet and put a tag on that little toe. I keep closing the eyelids and they won't stay shut and that blank look stares up. The sobs of the family can be heard all over. "Do you think Mom would want me to snip a lock of that blonde hair?' Why is there such pain, sickness and death?

I am stranded in this war zone. Who constructed an Emergency Room with no windows? We don't know the weather—unless we get a bunch of motor vehicle accident at once—20-minutes post rain. We don't even know if it's night or day without the military time. We opened our doors, turned on the lights and they have never been turned off—24-hours a day, 365 days a year and the sick never stop coming!

Chapter 6

Prescription for Suffering

The patient is 23-years old and his life is so screwed up on drugs. First, his aunt sends him in to get her fix and now he is lost to society. Mark Spellman, alias George Harris, alias Fred Harrison and I hear Dr. Remmy yelling, "I'm calling the FBI. I know you from Blastfield Community Hospital. Yes, I care about your health but I don't want to hear your bullshit stories. Now get up and get out!" The business office representative states, "Just give me your social security number." "No" he replied, "then you will know who I am."

Do you have any responsibility for your health? Remember, choices have consequences: you reap what you sew, more than you sew and in a different season. Ten or even 20-years of bad habits may feel good now, but bring horrendous results in the future.

The next patient is a 58-year old female who is SOB (short of breath) and on home oxygen therapy already. "I don't care what you do, just get me well." We all wish it was so simple. She has long stringy, gray hair and weight about 280. Her fingernails are long and yellow and she is reeking of cigarette smoke. She has diabetes and has blue toes. She has been in poor health for a long time and has a long list of chronic diseases. Her husband, like Jack Sprat—has poor color and is short of breath. He slipped her cigarettes in but said, "Just make her well." She needs a miracle!

The sixty year old male in bed 2 weight in at 380 pounds. He is covered with scaly lesions and said, "I don't believe you know why I'm here." "No sir, could you give me some symptoms, your complaint." He is coming off alcohol and Klonopin and drinks a 12 pack of beer a day—at least. He doesn't take "those prescriptions" and his girlfriend brings in the beer. He is on home oxygen. What a pitiful example of humanity!

This is a day of alcoholics and overdoses and a 16-year old experimenting with acid (LSD). It's only 10 AM and she has had 4 hits. She has a respiratory rate of 60 and is screaming—don't let me die—I'm not through with school. Why? "It expands you mind—kills you brand and heart." That's a very adult decision.

The 41-year old in bed nine said that she fell in the store. She is gray, clammy and pale. "I did drink 3 cases of beer a day but quit last week. " Well, the liver is gone. She has a hemoglobin of 5.3 (about 1/3 of normal) and has pain in the left hip. As the jeans are removed we see that the left leg is grossly bruised and edematous (there is probably 3 units of blood in there). She said, "I fell three days ago, too." There are large bruises on both arms and the left shoulder area. She might live to get out of the hospital—and to what future.

The forty-two year old is the second drunk of the day. He had a V-tach and cardiac arrest in October of last year. Now he is going into DT's—"hearing voices in the dark". He said he ran out of money and drink. Does he know that 1/3 of withdrawing alcoholics die?

I just peeked behind curtain 3. There is a tall black man lying very still with two large bore needles in his pecker. The Rx from his MD caused the erection and it wouldn't go away. Is this the treatment to return things to normal? Don't send that young female student in to assist with that procedure or she may die of shock.

An obese black man couldn't see the potty when he say down and didn't know he got lye (a caustic) on his scrotum. Now was this an outside toilet or what? Now there is an erosion on the skin and a constant dripping of blood. How do I hold pressure to control the bleeding on a squishy scrotum? Let's prepare for further treatment. Will cautery with silver nitrate do the trick or will he need sutures. Now we need creative thinking for the bandage.

The next patient has an altered mental state and states, "I am having a panic attack. I've always been messed up." as she is pacing and crying and tapping the wall with her head and pulling at the curtains to the next cubicle.

And then the young oriental male with a translators book in his hand. He is tying to describe his penile trauma.

What is a gyrating jaw? Crazy, just crazy—this place, this world. Nancy pours a hot cup of coffe and walks to the triage desk as three patients walk in the door. Could we have a rule that we need ten minutes between each emergency? The patient is lying on the seat in a motor home and blowing the horn for curb service. "Help me, don't touché me." Let's get a scoop stretcher to mover her. Will the dog bite if we go in for her?

What's wrong sir? He is one of the patients brought in on a backboard from the motor vehicle accident. Were you the driver of the vehicle? No, no I was just a bystander on the roadside and they grabbed me and put me on the backboard. Oh dear!

Oh no, the drunk guy with the finger injury answered to the family name of the man who just died. I told him his Father was dead and now everybody is confused.

"I am the richest man in Ireland." and stoned thinks Nancy. He is acting out or up but coworker Misty Skinny has him in a head lock. Last week she arm wrestled a man at the front desk for his gun and last month she ran after the psych patient and threw him to the ground and sat on him until help arrived.   I think she wanted to be a cop.

He has, "A vibrator in the rectum, don't remove it, just change the batteries." Tell me it's not so.

I only speak one language—hillbilly English—not Spanish, not Italian. Why do people think if they speak louder they will be understood better? Yes, we are all sick in the same language and body language can speak volumes.

Dr. Dooley is trying to gross out the new graduate nurses. He comes through the door shaking tootsie rolls out of the mountain dew he has poured into a bedpan and eating them. At least he is offering to share.

Example of a "drama code"—we often award daily winners. We do wimp awards too. It really is a way of dealing with life and death issues on a daily basis. Patsy walks through the ER door and collapses in a heap on the floor with rivers of tears and requests Demerol for her back pain, now! Patty to earth—Patty to earth—are you with us—the eyes are fluttering—Can you speak? Are you in pain? Wake up—the resident wants me to give Ativan IV for a seizure but I have never seen a seizure like this before.

Sometimes we just find strange things in strange places. The pelvic exam is interesting. There is a multicolored ring piercing through the labia—with a lock on it?

Tattoos may stay with you forever. He is 70 and his tool has a smiley face and it is till smiling. I have seen other strange tattoos—

are those vines leading to the root of all evil? Is that a banana on the coccyx? Should Tweety Bird be there?

He was working under the car and states he has an ear full of gasoline!

He has left flank pain and is pale and diaphoretic. Let's get the heavy jacket off. I think he has a kidney stone where it should not be. The pain has him on his knees.

How do we get the fish bone out of the throat? The last guy had a chunk of chicken obstructing the esophagus after choking and having the Heimlich maneuver done. He couldn't even swallow his saliva. Meat tenderizer used to be the drug of choice until someone got a hole in the esophagus. At least he is alive and breathing.

He was wearing tight leopard skin colored thongs?

Susan has a new found breast lump. What will her future hold?

The skinny little lady keeps moving something around in her mouth. What is wrong with her? "Oh, that's just my gum. I like to chew two packs at once."

The 23-year old male refused to remove the earring for the cervical spine x-ray and now objects to returning because an accurate reading couldn't be given because of a view blocked by an earring. We don't keep the earring. You can put it right back in. Now, many people don't want to remove their wedding rings.

Why is the curtain fluttering in Room 8? What are you doing? "I'm blowing but the only problem is if I close my eyes I go off to Jupiter." It is not hard to figure out what kind of treatment he needs.

Triage means to sort out. Now, I will try to sort out this patient. Male or female? I see chest hair, shave the face, a man's shirt, pony tail, aha—false fingernails. It's a girl, I think. Should I ask about the last menstrual cycle?

The little 76-year old female is sobbing her heart out. She said, "My husband died in November and my son moved in and they won't let me come out of the bedroom. My son said if my wife leaves me you don't know what I'll do to you." She said, "Don't admit me to the crazy ward—he will only use that to take my house."

It would be impossible to produce this scene again: We have a screaming child being sutured and the 93-year old in the next cubicle is yelling—"Oh stop, what are they doing? Why don't they give her a shot? Isable get my breakfast." As the ventilators are alarming and the newly diagnosed bran tumor starts another seizure and I am trying to mix multiple drips for the acute MI patient. "Donuts, anyone?" asked the Drug Representative. We have one PID (pelvic inflammatory disease), two kidney stones (who desperately need pain medicine) and a child with hives—someone get the epinephrine—for starters.

Don't lay my shoes on top of my panties—pointing to the Kleenex. Put those cigarettes away and don't smoke—as the young MD, playboy Shirella does a rectal exam. She screams—"You have no respect for your elders."

The frizzy little blonde has a body to kill for and has a burning pain in her left arm. "I did acid tonight."

Lets get a handle—which Doc is coming in? That 265 pound gentleman is short of breath because of one short episode of bleach inhalation? Could the extra weight and smoking two packs a day have anything to do with it?

Is there an HIV epidemic here? The guy in the waiting room asked, "How's my lover?" When he hit that car I peed my pants—as the wet splotches show on his slacks.

The skinny 24-year old bleach blonde said he took $100 worth of cocaine, plus valium and alcohol. "I'm HIV positive and I'm trying to kill myself." He is loud and abusive and fighting. He took Nurse Kathy to the floor, spit on her and slapped and sat on her before we got to them. No drama here. Maybe Oprah would like to film today.

Her face is the color of my favorite blue sweater—Emergency? Let's put her in a treatment room. The vital signs are good—the oxygen saturation is good. "Do you feel short of breath?" "No, I sprained my ankle. My face always looks like this. As a young lady I used too much nasal spray." That is blue face! Reminds me of a young lady who have very blue legs and we finally diagnosed new jeans with fading colors.

This patient has a note: "I need a trash" He's talking and waling has good color, but I know looks can be deceiving. What is this about?

He has fish meat belly color—interpreted, very pale—and a scalp laceration with a hole big enough for my hand and he just wants a dressing put on!

The blonde beauty has bikinis, fish net hose, spiked heels, bracelets, earrings, bright lipstick and male parts. Don't send her— him to the semi-private room with Sarah.

Here comes a PID shuffle (pelvic inflammatory disease). I can recognize that walk a mile away: belly pain, vaginal discharge and wearing house slippers that she can't raise off the floor.

New Nurse Hannah took the k-lyte to frail little Mr. jones to swallow. "The pill is larger than a quarter and should be dissolved in a cup of water." ***We need to be wary and be our own health care advocates as nurses and doctors are just people and make mistakes too.*** At least this mistake wouldn't have been life threatening.

Jim has a bucket full of medicine. "I don't need to see a Doctor, I just need a few prescription refills. " And what conditions or illnesses and which Dr. initiated this long list of medicines?

A sure sign of old age—worried about the bowels instead of sex!

He said—she said—who is telling this story. He said she fell against the picnic table to receive the large circular laceration around the neck. She said he cut her.

We need monitored bathrooms—he later tells us that a pricked finger adds enough blood to the urine to feign a good kidney stone and receive lots of pain medication.

Is your teaching understood? A clean catch urine means to put some urine in the bottle in the middle of the stream after cleansing yourself with the wipes. I gave her the wipes and urine cup and told her to clean herself from front to back and to start to urinate in the toilet and catch some urine in the cup. She wiped her arms and hands with the wipes, put them in the cup and peed in the toilet. Go figure how we can tell if she has a urinary tract problem.

The rich and famous get sick too—You look just like Joan London—I am Joan London. I know that you are the President's wife, but we still need a set of vital signs and a chart to check your eye. No, we won't mess up your hair.

An early morning shower I didn't want. As I am helping the very pregnant young lady on the exam table—her bag of waters rupture—but we don't here—do deliveries here—usually.

Yes, this ER could run more efficiently. The secretaries spent all morning entering data into the computer and it didn't store any of it. At the same time the nurses ran to the lab, answered phones and processed the Dr.'S orders. Keeping supplies at hand is a big job. When I reach for an oxygen mask or a defibrillator pad it could be a life and death situation.

And Harry thought his treatment was fine until he went to pay the bill. $846 for that tiny kidney stone! The revealing IVP was $344. "I didn't need that." Now that you are pain free and the stone has passed you don't.

The little guy in room 3 is whiter than the sheet. His hemoglobin of 6 makes him a few pints low. We need tow IV lines now and type and cross match for 2 units of PRBC's. He will need to be watched closely for an adverse blood reaction.

Overdose X 2 and it isn't even "Happy New Year" or anything. One holdover has a blood alcohol of 347 (three and 1/2 times legally drunk) and the other gal need her stomach pumped and the charcoal treatment. Was the overdoes accidental or intentional?

Here is one big gat knee attached to a gentleman who has a badly discolored foot and impaired circulation. Those who need to be here often wait too long and some are such frequent fliers that we know their life story. That lady had 17 ER visits last month. How may Doctors and how many drugs? We can't refuse to see anyone.

I know sitting in the waiting room with bunch of sick people is hard. Grandma, the child has nausea and vomiting. Please don't give her milk and cookies. Just let the gut rest and she could have something surgical. Yes, we know she is very sick and yes we will see her.

Some situations will just never be resolved. A male Dr. needs a female nurse to do a pelvic exam, but a female nurse can put a catheter in a male patient all by herself.

Many of our patients don't even know what the medicines are for much less what kind of side effects to watch for. People, you must be your own advocate and educate yourself and coordinate your care, especially if you have more than one Doctor.

Sweet little Libby in bed 2 feels so much better after the oxygen and nitroglycerin that she want to go home now. Wrong Libby, that means that it is probably heart related and that you need to stay awhile.

Thin beautiful Carmen says her heart is hurting. She thinks we can fix an emotionally broken heart. Her husband filed for divorce three weeks ago and she has lost 15 pounds and can't sleep or eat. Well, there are medicines to help in the interim and counseling; but grief takes time.

The guy with diarrhea won't undress. He said he doesn't need much—just a few pills to plug him up. Sorry George—we can't work on the motor unless we look under the hood. We need to fell the belly.

A lot of people are very sick with this virus that is making it's rounds. The bugs seem to hit harder and last longer. There are high fevers, chest congestions and body aches. The young men all think they are dying. And there are no miracle cures—rest, fluids, time and treat the symptoms with lots of rest. *No one wants to take the time to care for their health—and what else is there?*

Gee, who ever came up with the 9 AM to 9 PM shift? I know it covers deficits but that pretty much takes my day. I do moonlight with house, husband and children. Three of us older gals are called the "three stooges" and we must be really dedicated or stupid to work this shift and this hard. No on seems to be grateful, "Are you going to stick me again?" "This old bed isn't very comfortable." "What do you mean, I can't get up to the bathroom?" (She has no blood pressure." "Do you really only have one Doctor here?" "I just need a stitch or two." And the next patient needs a new heart and Johnny just need a Dr. to look at his throat and Eleanor just needs the Doctor to peep in and order a shot for her migraine. She knows what she always gets and on and on it goes. The health care benefits are good and all the viruses, staph, TB, hepatitis and HIV come free— especially the psychoses after dealing with the sick and dying all these years. My co-worked Brandon moves so slow the patient will be dead before he gets the IV in and gives her some atropine to speed the heart rate. Then he pushes the drug so fast she become instantly flushed and sits straight up in the bed and says, "Oh, Oh, Oh!" I guess he will remember from the day forward to push atropine a little slower. I will too.

It's lunch time and I crawl into the break room to enjoy a few minutes of peace and rest. The cleaning person is running the sweeper and cleaning the John with bleach. And someone ate my lunch from the fridge and I don't have a penny in my pocket. Was the cookie monster here or just a hungry soul? Will a stale cracker get me through the afternoon? Do I want mustard or ketchup on the cracker?

The Doctor is just four procedures behind. We have one breast that needs an incision and drainage (nipple right got infected); two wounds (one a rather large ugly laceration) to repair; and one knee to tap to remove excess fluid. Yes, these parts are all attached to nice people.

We do get some strange things in strange places in this ER. Dr. Raynard just delivered a naval orange from the vagina of a 19-year old. He had to peel it inside to remove it. And then the young lady who came in to have a pickle removed. That young man is having a problem. He said he put a birthday candle in his penis and lit the candle and it melted. I don't think the procedure books tell how to remove melted wax from there. Patient Fred said he was expelling gas and didn't want to be embarrassed at a party so he plugged the rectum with a pacifier. Now he can't retrieve the pacifier. I'm embarrassed!

Patient Josephine has a snicker bar lost in the vagina. Why didn't she take the wrapper off? I know it's Christmas but who would have thought of finding Santa Claus there! It is a stuffed toy alright.

Call the Sheriff, this lady is after the patient with a gunshot wound to the legs. She said he wasn't getting a freebee. She's a prostitute and here to collect.

That looks like a metal cross sticking out of the top of his head. His co-workers said they trimmed the end of a cross bar used to reinforce sidewalks to get him here. He's alert and the bleeding is controlled. Ho far could this thing be embedded?

Now the moral of this story s not to store knitting needles by the closed door. Patient Mary Lou said she climbed on a stool to look for a purse and fell backwards and the knitting needle caught her in the behind. One surgical procedure and punctured bowel later the needle is removed.

I would call this strange things in strange places. Dr. Stumps who weights in at 240 pounds delivered a baby in the back seat of a Volkswagen Bug. Close quarters—it's a boy. Curb service—we are a full service hospital.

Dr. Dyphers who loves to play pranks was called back to the morgue to pronounce a patient (a grossly, badly, mangled) body bag kind of patient to be pronounced. When he pulled the top off the morgue cart Nurse Karen grabbed him around the neck. He screamed and ran three blocks outside the hospital before we caught him. Do you think he will put KY jelly on the toilet seat tomorrow? Nancy thinks, we have to laugh to keep from crying with so much sickness, tragedy and sorrow.

The young man in cubicle three has a carrot in the rectum. He said he was cooking naked and the pressure cooker exploded. That's his story and he is sticking to it!

There's a telephone call for Mike. Le me check the waiting area outside. There's only a lovely lady there wearing a sexy dress, black net hose, spiked hells and perfect makeup. Yes, my name's Mike. OK.

The patient in bed 8 is constipated, "no BM for 9-days." I can see enemas in her forecast—high, hot and a hell of a lot. I need help carrying the potty chair. She delivered a six pound stool. Would that ever make you miserable. Oh gee, will the department smell like this all day?

There's been an auto accident and we are getting four patients. All are back boarded with cervical collars in place. Does everybody have neck pain? They are all Haitian and can't seem to talk when they get sick. Only moans are heard and mumbling to their Voodoo kind of medicine. They must be afraid. The highway patrol is here and two of those patients were seen jumping into the car after it wrecked. You may get up and leave or go to jail. Miracle healing— they're up and out of here.

Nancy thinks she has seen everything until she looks at the chest x-ray hanging on the view box. What are those round things? He swallowed condoms filled with heroin. One ruptured and he is ill with heroin poisoning.

The young man said he shot himself with a BB gun and has a headache. He should be thankful for the BB gun because his CT shows a huge brain tumor totally unrelated. He will be sent to the neurosurgeon and what will his prognosis be?

We have a family who fell off a boat. The boat tilted and Dad had near drowning. The papa weighs 500 pounds, Momma 400 and little boy 300 pounds. Wonder why the boat tilted? Maybe this will be that teachable moment—to chooses better eating habits and better health.

What we do to ourselves with what we stuff into our mouths. We end up with rolls of fat to be pushed, pulled, belted and strapped. Doing a pelvic exam on patient Claire took three people; tow to hold back the rolls of fat so the Doctor could find the vagina. With gallbladder disease and diabetes, heart and breathing problems—just add cigarettes and you can work up to some wonderful, awful health problems. Then the nurses gather in the lounge for a smoke and fresh pastry. We never seem to learn!

Nancy sits in triage to sort out the illnesses and takes a history. How much do you weigh—the answer is often, "too much" or they say 140 and look like 180. The men always seem to know what they weigh. And the gray color and wrinkles often tell the smoking story or history. They often say, "I don't smoke—I quit yesterday."

Nancy was having lunch in the lounge watching the new flash—wanted for murder—oh my, that's the patient in bed 1.

Triage can be an interesting place—the man looks over the desk at Nancy and said, "If you laugh, I'll kill you." He looks the part—wearing a trench coat and his hat pulled down over his eyes. He has his tender part stuck in his zipper.

Talking about strange things in strange places. A tour bus pulled into the ER lot. One of the students—a young little thing—delivered a baby into the toilet on the bus. Once the paramedics removed the tank and retried the child, he was dead. A beautiful little 6 pound angel. The student is still saying that she couldn't be pregnant.

Diagnosis per Dr. Jays, "The Vines." Older medicine used a pessary ring to help hold a woman's uterus in place when prolapse was a problem but patient Ella Mae used a potato slice. The potato sprouted—thus, "The Vines". I couldn't even dream this stuff up.

George eats Barbie dolls for his kicks. He said he eats the heads (19 so far) because he likes the way it feels when they pass. To each his own, I guess but I think we need a psychiatric consult. I think he's missing a few peas in the pod; two sandwiches short of a picnic or a few marbles short of a bag. It sure is hard to keep a straight face sometimes.

Chapter 7

The Battle Axe

This is one of those ER days for my memories. Code yellow repeated four times over the intercom. Now let's see—What does that mean? Code blue is an adult who is not breathing or has no heartbeat; code 45-a pediatric patient with no breathing or heartbeat. Code brown is a weather warning—like a tornado. Code yellow is a bomb threat—Where di the message come from? There are two bombs placed in the hospital to go off within the hour! Can I go home now? The place is chaotic already because we are trying bedside registration in order to expedite care—to compete in today's managed contracts. Yes, the hospital needs ER patients as they compromise about 98% of hospital admissions.

A little fossil from your local nursing home is yelling and trying to climb out of bed. A hysterical black lady with chest pain is crying loudly. There is a drug we use that sedates you and you don't care if you hurt—bring on the Inapsine. One little man with an old CVA—stroke—is hollering—don't stick me again. He is pale and diaphoretic with his nervous wife hovering and patting. He had a hypertensive crisis during his physical therapy. And the little old Dr. in bed 4—who was offended because the triage didn't address him as Dr. is trying to direct his own care—just a shot of Lasix please (that's a diuretic) and he will go on about his business. If his congestive heart failure gets worse he will come back—if he lives long enough.

Did Nancy Nurse know that she would need continuing education credits forever to use her R.N. License? When we were in school did we even think to envision that changing technology; computers and drug dispensing machines? Where were the classes on endurance, patience, communications and survival? Per the last patient—"I got a shot right in the middle, the behind. The nurse hit the sciatic nerve and I can't see, can't work and need pain medicine." The tech, Kevin did a blood draw and after sticking the patient and pulling the needle out scraped it across the arm and when the patient moved it ripped a hole in the arm. Oh, no we will never hear the end of this. We really are human and errors do occur.

The patients expect so much. The prescription hasn't been filled. The fever hasn't been treated. Do we have a magical cur—zap, zap, zap? Does this cost anything? No, all medical services is free— for some and the rest of us just pay higher premiums.

A typical ER day. It started in slow and then started to grow. By 11 AM we felt like we had been bombed. I have walked thousands of miles, treated and cared for thousands of people, helped make millionaires of several MD's and served humanity—and lost my sanity.

Dr. Bald is leaning sideways in his chair. With slurred speech he says, "I think I'm having a stroke." Wait, our only ER Doc can't stroke, our house is full of sick people! Bring a stretcher to the Doctors desk. Call for an EKG, lets start an IV line, get him on a monitor and oxygen. Arrange for a CT of his head and call the hospital supervisor—can you find me a Doctor—somewhere? How long will it take to helicopter someone in from another location?

Am I ever glad to be going home! There are four new ambulances, a fire truck and a taxi in the ER entrance. One stretcher just unloaded a patient with CPR in progress. And the patient I just left complained that "her breast was exposed". We were just trying to do an EKG.

AD—Enticement for ER nursing: There is lots of drama, excitement, first hand news, Code 90's, handsome Doctors, bomb threats and exposure to every ailment known to man. You can serve humanity, develop insanity and create foot, leg and back problems. You can work long tedious hours, including call ins at 3 AM and weekends and always spend the holidays with your co-workers. There is a free fitness training which includes pushing, pulling, tugging and lifting. You can explore great culinary delights which might include cold coffee, missed meals, bites of pizza between the bedpan changes and learning to eat dinner in three minutes flat. But don't bring food to the nursing station or the little redheaded manager may stomp on your grave.

This has been a grueling day with patients in wheelchairs, on stretchers, coming in ambulances and screaming with pain. They are falling from ladders, pregnant and bleeding and wanting to know if they can be tested for venereal disease. Those telephones never stop ringing, some very vague and frightened callers to the Doctor who wants the name of the patient admitted to his service during the night. He "don't have a clue of the name or problem."

"The sick you will always have with you—or was that the poor?" thinks Nancy as the med com goes off. It has been 20 minutes since the rain started and we are getting five patients from an MVA. Will people never learn to slow down in the rain? OK, so another cup of coffee gets cold. Actually, I like cold coffee—it grows on you after so many years. Sarcasm, irritability, hardcore and it's not just me but I need out of this ER. I'm sick of sick people, well not all of them but the rudeness of some makes it difficult. The manic depressive patients from yesterday looked at me and said, "You are smacking your lips and I don't like you." and I was trying so hard to help her.

Nobody likes to be stuck with a needle and I have stuck 9 million, 3 hundred and two. Not one of them liked it. What a thankless job and heaven forbid that a little bruise appears. "Look what that nurse did to me!" Never mind the fact that she was trying to save your life.

Mr. Jones had vascular surgery six days ago and was just discharged from the hospital this morning. The suture lines are healing beautifully and the surgery went great. Wonder why he's dead? And Leon Lasko is only 45 and on vacation with his family. His wife said he is a healthy as a horse. The chart reads: healthy 45-year old male—expired at 0830. Professional caring people would not talk about skid marks in undies or body part sizes or barter for a dead man's shoes, would they?

Remember the ABC's of airway, breathing and circulation. I can't breathe because of offensive odors in the air and I must circulate to another room. Where's my purse size right guard I usually reserve for sneaker feet odors? Code 90, zap, zap, ZAP: little zap, big zap—everybody zap. How many miles of walking, how many patients? Now may I retire, please!

These patients expect so much—miracle cures and they don't even fill the prescriptions or don't treat the fever, zap, zap, zap. Dr. B is the patient in bed 5 ready for discharge? The urinalysis is back but now I have an order for a chest x-ray as the patient impatiently paces up and down by the nurses station watching her watch.

No more excitement please—that is just a board nailed to her hand and she can't straighten the fingers—that nail gun must have been vicious. That is just a concrete saw tooth embedded in the leg— not too much blood, it doesn't look too deep and the next patient has a screw in the neck—stranger things have happened, get a screwdriver.

Why are the paramedics wearing masks, "fungus among us", funky mungus—just BO? Just take a deep breath before you go behind the curtain. I think he also has a little does of polar bear syndrome—old booze. Patient Fred just wants a prescription for Nicorette gum, he said he has chewed it for 9-years; so much for short term help to stop the smoking habit. Here comes the little redhead: alias ER Nurse Manager, "Who cut all the straps off the backboards—at $68 a strap?" Could it be the med student who still thinks everything is urgent? Did you ever wonder why your blood pressure is checked three times—maybe the student just doesn't know how to take a blood pressure, she got a wrong reading or you're almost dead. The nurse may be old and her hearing is going fast or the sphygmomanometer is broken or you are too fat or too skinny for the cuff.

The Doctor wants a gas mask—what's coming in—as I round the corner to the patient bathroom the call light comes on. Which patient has that explosive episode? I see the other patients covering their heads, going down the hall from x-ray: visitors gagging and the physician assistant laughing his head off and Dr. Dooley staggering down the hall. Just wait until that door opens a little more. I'm glad my sinuses are closed today. In fact, I felt so bad I traded Dr. Dooley a piece of cake to take a power nap on his bed at lunch time. Who mad the stupid rule of no sleeping at work. A time and cot would help the quality of care.

The new year started with so many tragedies, I revised my resolutions for the year:

1. More chocolate—less beans.

2. Wear only t-back undies—why not let it all hang out.

3. More hugs and power naps.

4. Go barefoot and nude more often.

5. Say yes.

6. Buy me flowers once a month.

7. Don't spend all of off days mopping floors.

8. Laugh more.

Life is short and then you die.

That ER—A mother runs in the door and throws her baby at me. "He stopped breathing!" Those big eyes looking up at me. She is screaming—don't let him die. He's 9-months old. Why, Why? Labs of course. They can't be Hemolyzed—a bag for urine. Vital signs—as an ER full of patients and visitors are pacing, waiting, watching My daughter just needs her IV out. Yes, I know—but this baby is barely breathing. The Mother—but I have to get back to work. Can any job be so important?

That is 71-year old Gertrude in Bed 5—Dr. Dooley says with a nice shape—for an old lady. Her husband has walked in and out of the treatment room about 900 times. Please stay in the treatment room—others like their privacy and you don't want to be run over by a nurse.

Draw labs—
I hate this."—Don't' hurt me. She had an abortion 3-weeks ago and now a mass? Is it an abscess, an ectopic pregnancy or tumor?

This is bike week and full moon. Will the ER personnel survive? Another vomiting baby. His mother said he won't take his medicine or drink. Please offer a bottle, a syringe, a cup of ice cubes. He'll get the message. I wish we could give a prescription for parenting classes. The basic skills often seem to be sadly lacking.

Is someone's car horn stuck in the parking lot or does someone need curb service? I'm compelled to go look—the last horn that blew was a tour bus with a dead patient sitting in the back. Oh I shouldn't have looked, her comes Bill Wilson, he has the story down pat—chest pressure with nausea, sweaty and I need morphine now. He is a sad cardiac cripple. I just want to put his head in a bag with a twist tie on top. I am too tired, my tootsies ache and I know three more ambulances are coming in and we have no empty spaces to put them. The patient in room 1 needs more pain medicine for his kidney stone and patient in bed 2 needs water and bed 3 need help up to the bathroom. And the patient in bed 4 wants someone to make a couple of phone calls for him. The sick you will always have with you, zap, zap, zap.

Nancy thinks my big toe is really sore, may Dr. Dooley will take a look before I go home. No such luck as there are 9 patients on the deck yet to come in to treatment rooms and that waiting room looks like a rock concert is happening. They are on the floors, hanging off the tables, Kentucky fried anyone? They could have brought enough to feed us all. Just mingle all those viruses and ailments together and we will have an abundance of patients next week. The police are giving tickets for the auto accidents and all aches have become more pronounced.

Dr. Pickle, she's ruped. (The patient who needs the pelvic has her feet in the stirrups.) I hope the light don't burn a hole in it before he gets to the pelvic exam. He should try that position for a bit. Dr. Pickle, "Okay, who's going to slide down with me?"—assist with the pelvic exam. And there come Dr. Bee from the patient in bed 6 after saying to the patient, you're going to be okay. To Nancy, "he's going to die, he's going down the tubes, circling the drain—big time." There has to be some mechanism to cope daily with the sick and dying.

Nurse Patty is trying to take a knife from the patient in bed 3. Patty please give it up—better him than you. Those young nurses come out of school ready to save the world. Patty, we do make a difference, many are helped and many cured. Death is a real part of life and a path which we all must take some day.

I spend so much time with these Doctors, pick up after them, assist, help, maybe advise; like a wife—just no fringe benefits. MD Ricky is so cute—I don't know what came over me, when he walked by I tweeked his buns. If I were a little younger and not so tired! Don't let the little redhead know there is life outside the ER. I'm scheduled to worked 10 days out of 11.

It's 5:30 AM and Nancy just can't seem to get out of the bed. 5:30 didn't used to seem so early. This toe really hurts. Call 911, Nancy Nurse is ill—"I am so sick." But wait, I must take a shower, they do notice if you have on clean underwear. Even when they see you in the grocery store next month they remember the skids. The tattoo is clear on my chest, NO CODE—do not resuscitate, please don't plug me to machine to take my life savings to pay hospital bills; don't send me to the nursing home t lie in a contracted heap with bedsores on my bottom and smelling like old pee. No heroic measures please. Death is part of life!

Nancy arrives in her own ER via ambulance: old and gray with a high fever and chills. Is she hallucinating? She said the computers are after her—she couldn't make all the entries and do patient care too. She said the paper work isn't finished. Rooms are full of unfinished paperwork. The lawyers are on the doorstep—not documented, not done—not documented—not done. She has been exposed to TB, Aids, Meningitis and every other disease known to man. That right great toe looks bad with red streaks going up the leg and gross edema. Could she be septic? What does she man, she can't go—her shoes aren't polished. The dog tag is tangled in her stethoscope. Zap, zap, zap. "Patient care is priority"—nurse fluff my pillow, I need water, when will my room be ready. I know I can't to my room yet because bed 15 is not clean in the computer. How did we ever take care of patients without computers? Nancy is singing Christmas carols in July; 6 patients with nausea, vomiting and diarrhea, 5 chest pains, 4 sore throats, 3 facial lacerations, 2 broken arms and 1 drug seeker.

Hey, can you jump start my heart? No warranties, no major tune-ups, just emergency repairs. Code 90, Code 90, Code 90 comes three times over the speaker. When Nancy's chest is exposed there is a tattoo that reads, "Do Not Resuscitate". Nice boobs, she always said she would never wear old lady underwear—hot pink striped bikinis with lace. Zap, Zap, Zap. No response. Nancy goes the way of all the world. Only now she has rest and peace. "Good-bye Nancy." Dr. be declares, "I will never forget you!"

Words for thought

Oriented x 3—knows name, place and day

Turtle—the patient died—only the shell here

Toe tag—body bags

Death-dying-grief

Birth-smiles—home remedies

Miscarriage—sadness

Suffering, pain, mending, repairing

Prescriptions—healing

Normal blood sugar 80-110

Master of your own destiny—eat, drink and die

Computer rule.

Frequent flyer—comes to the ER often

Drama code—really plays u a minor complaint

Code brown—smelly poop

Money, twat—vaginal area

Put you money—bet on the blood alcohol level

Code 90—adult cardiac or respiratory arrest

Code 45—Pediatric cardiac or respirator arrest

Circling the drain—CTD—fixing to die

Positive suitcase sign—wants to be admitted to the hospital

Sever sign—Seaver's is the local mortuary

Zap, Zap, Zap, Everybody zap, little zap, big zap—sequence of shocks and drugs in a Code 90 situation.

Tidbits of thought picked up from fellow ER nurses.

You might be an ER nurse if you have your weekends planned for a year in advance—because you know you are working every other weekend.

You automatically assume the patient is a drug seeker when presented with a complaint of migraine, low back pain or chronic myalgia.

You disbelieve 90% of what you are told and 75% of what you see.

You believe in aerial spraying of Prozac...

You have the bladder capacity of five people...

You refer to vegetable and not talking about a food group.

You plan you dinner break while lavaging an overdosed patient.

Your idea of comforting a child includes placing them in a papoose restraint.

You have discovered a new condition that you call, "hypo-Xanax-emia".

You believe that unspeakable evils will befall you if the phrase, "wow, it's really quiet" is uttered.

You say to yourself, "great veins" when looking at complete strangers.

You have ever referred to someone's death as a "Celestial Transfer".

You have seen a diagnosis—lost condom.

You refer to someone in severe respiratory distress as a "smurf".

You idea of a good time is dueling shock rooms.

You believe that "too stupid to live" should be a diagnosis.

You have ever had a patient look you straight in the eye and say, "I have no idea how that got stuck in there."

You have ever had to leave a patient's room before you begin to laugh uncontrollably.

You have ever wanted to reply "yes" when someone call the ER to ask "Is my husband, wife, mother, brother, friend there?" Yes and his girlfriend, etc.

Your favorite hallucinogen is exhaustion.

You think caffeine should be available in IV form.

You have restrained someone and it was not a sexual experience.

You have witnessed the charge nurse muttering down the hallway, "Who's in charge of this mess anyway?"

You believe the waiting room should be equipped with a Valium fountain.

You believe a "Supreme Being consult" is your patient's only hope.

You want the lab to order a "dumb shit profile".

You have been exposed to so may x-rays that you consider radiation a form of birth control.

You know your patient is demonically possessed.

You have ever had a patient say, "But I'm not pregnant, I can't be pregnant. How can I be having a baby?"

You carry your own set of keys to the "Leathers".

Your idea of gambling is a blood alcohol level instead of a football pool.

Your feet are slightly flatter and tougher than Fred Flintstone's.

Your immune system is so well developed that it has been known to attack squirrels in the backyard.

You have recurring nightmares about being knocked to the floor and run over by a portable x-ray machine.

Your shoes have been seized and quarantined by the Center for Disease Control.

You're able to tell the difference between a medical order and chicken scratching.

Your idea of thawing the holiday turkey consists of an IV and warmed saline.

You start an IV on the pumpkin in the backyard to enhance it's growth.

.